Activity Book
HUMAN BIOLOGY AND HEALTH

Prentice Hall
Englewood Cliffs, New Jersey
Needham, Massachusetts

Activity Book

PRENTICE HALL SCIENCE
Human Biology and Health

© 1994, 1993 by Prentice-Hall, Inc., Englewood Cliffs, New Jersey 07632.
All rights reserved. Student worksheets may be duplicated for classroom
use, the number not to exceed the number of students in each class.
Printed in the United States of America.

ISBN 0-13-225517-0

    6 7 8 9 10   97 96 95

Prentice Hall
A Division of Simon & Schuster
Englewood Cliffs, New Jersey 07632

Contents

- **To the Teacher** H4

- **CHAPTER 1 ■ The Human Body** H7

- **CHAPTER 2 ■ Skeletal and Muscular Systems** H25

- **CHAPTER 3 ■ Digestive System** H41

- **CHAPTER 4 ■ Circulatory System** H69

- **CHAPTER 5 ■ Respiratory and Excretory Systems** H89

- **CHAPTER 6 ■ Nervous and Endocrine Systems** H107

- **CHAPTER 7 ■ Reproduction and Development** H133

- **CHAPTER 8 ■ Immune System** H149

- **CHAPTER 9 ■ Alcohol, Tobacco, and Drugs** H171

- **Science Reading Skills Worksheets** H187

- **Activity Bank** H203

To the Teacher

The materials in the *Activity Book* are designed to assist you in teaching the *Prentice Hall Science* program. These materials will be especially helpful to you in accommodating a wide range of student ability levels. In particular, the activities have been designed to reinforce and extend a variety of science skills and to encourage critical thinking, problem solving, and discovery learning. The highly visual format of many activities heightens student interest and enthusiasm.

All the materials in the *Activity Book* have been developed to facilitate student comprehension of, and interest in, science. Pages intended for student use may be made into overhead transparencies and masters or used as photocopy originals. The reproducible format allows you to have these items easily available in the quantity you need. All appropriate answers to questions and activities, are found at the end of each section in a convenient Answer Key.

CHAPTER MATERIALS

In order to stimulate and increase student interest, the *Activity Book* includes a wide variety of activities and worksheets. All the activities and worksheets are correlated to individual chapters in the student textbook.

Table of Contents

Each set of chapter materials begins with a Table of Contents that lists every component for the chapter and the page number on which it begins. The Table of Contents also lists the number of the page on which the Answer Key for the chapter activities and worksheets begins. In addition, the Table of Contents page for each chapter has a shaded bar running along the edge of the page. This shading will enable you to easily spot where a new set of chapter materials begins.

Whenever an activity might be considered a problem-solving or discovery-learning activity, it is so marked on the Contents page. In addition, each activity that can be used for cooperative-learning groups has an asterisk beside it on the Contents page.

First in the chapter materials is a Chapter Discovery. The Chapter Discovery is best used prior to students reading the chapter. It will enable students to discover for themselves some of the scientific concepts discussed within the chapter. Because of their highly visual design, simplicity, and hands-on approach to discovery learning, the Discovery Activities are particularly appropriate for ESL students, in a cooperative-learning setting.

Chapter Activities

Chapter activities are especially visual, often asking students to draw conclusions from diagrams, graphs, tables, and other forms of data. Many chapter activities enable the student to employ problem-solving and critical-thinking skills. Others allow the student to utilize a discovery-learning approach to the topics covered in the chapter. In addition, most chapter activities are appropriate for cooperative-learning groups.

Laboratory Investigation Worksheet

Each chapter of the textbook contains a full-page Laboratory Investigation. A Laboratory Investigation worksheet in each set of chapter materials repeats the textbook Laboratory Investigation and provides formatted space for writing observations and conclusions. Students are aided by a formatted worksheet, and teachers can easily evaluate and grade students' results and progress. Answers to the Laboratory Investigation are provided in the Answer Key following the chapter materials, as well as in the Annotated Teacher's Edition of the textbook.

Answer Key

At the end of each set of chapter materials is an Answer Key for all activities and worksheets in the chapter.

SCIENCE READING SKILLS

Each textbook in *Prentice Hall Science* includes a special feature called the Science Gazette. Each gazette contains three articles.

The first article in every Science Gazette—called Adventures in Science—describes a particular discovery, innovation, or field of research of a scientist or group of scientists. Some of the scientists profiled in Adventures in Science are well known; others are not yet famous but have made significant contributions to the world of science. These articles provide students with firsthand knowledge about how scientists work and think, and give some insight into the scientists' personal lives as well.

Issues in Science is the second article in every gazette. This article provides a nonbiased description of a specific area of science in which various members of the scientific community or the population at large hold diverging opinions. Issues in Science articles introduce students to some of the "controversies" raging in science at the present time. While many of these issues are debated strictly in scientific terms, others involve social issues that pertain to science as well.

The third article in every Science Gazette is called Futures in Science. The setting of each Futures in Science article is some 15 to 150 years in the future and describes some of the advances people may encounter as science progresses through the years. However, these articles cannot be considered "science fiction," as they are all extrapolations of current scientific research.

The Science Gazette articles can be powerful motivators in developing an interest in science. However, they have been written with a second purpose in mind. These articles can be used as science readers. As such, they will both reinforce and enrich your students' ability to read scientific material. In order to better assess the science reading skills of your students, this *Activity Book* contains a variety of science reading activities based on the gazette articles. Each gazette article has an activity that can be distributed to students in order to evaluate their science reading skills.

There are a variety of science reading skills included in this *Activity Book*. These skills include Finding the Main Idea, Previewing, Critical Reading, Making Predictions, Outlining, Using Context Clues, and Making Inferences. These basic study skills are essential in understanding the content of all subject matter, and they can be particularly useful in the comprehension of science materials. Mastering such study skills can help students to study, learn new vocabulary terms, and understand information found in their textbooks.

ACTIVITY BANK

A special feature called the Activity Bank ends each textbook in *Prentice Hall Science*. The Activity Bank is a compilation of hands-on activities designed to reinforce and extend the science concepts developed in the textbook. Each activity that appears in the Activity Bank section of the textbook is reproduced here as a worksheet with space for recording observations and conclusions. Also included are additional activities in the form of worksheets. An Answer Key for all the activities is given. The Activity Bank activities provide opportunities to meet the diverse abilities and interests of students; to encourage problem solving, critical thinking, and discovery learning; to involve students more actively in the learning experience; and to address the need for ESL strategies and cooperative learning.

Contents

CHAPTER 1 ■ The Human Body

Chapter Discovery
*Discovering Levels of Organization.........................H9

Chapter Activities
*Activity: What Is a Tissue?................................H13
*Discovery Activity: Medical Terminology Made Easy...........H17
Discovery Activity: Describing Skin.........................H19

Laboratory Investigation Worksheet
Looking at Human Cheek Cells................................H21
    (**Note:** *This investigation is found on page H24 of the student textbook.*)

Answer Key..H23

*Appropriate for cooperative learning

Name _____ Class _____ Date _____

Chapter Discovery CHAPTER **1**

The Human Body

Discovering Levels of Organization

Background Information

In multicellular living things, the work of keeping the living thing alive is divided among different parts of the body. Each part has a specific job to do. And as the part does its specific job, it works in harmony with all the other parts of the body.

The groupings of these specific parts within most multicellular living things are called levels of organization. The levels of organization in a multicellular living thing include cells, tissues, organs, and organ systems. In this activity you will discover how cells, tissues, organs, and organ systems are organized.

Materials
construction paper: yellow, green, blue, red
scissors
tracing paper
glue
marking pen

Procedure
1. Trace the outline of the triangle shown in Figure 1 on tracing paper. Using scissors, cut out the triangle from the tracing paper.

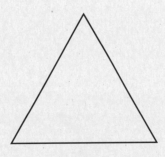

Figure 1

2. Use the tracing paper triangle to make four identical triangles on yellow construction paper. Cut out the four yellow triangles. Label each triangle with the word CELL.
3. Place the four yellow triangles on the sheet of green construction paper so that they together form a larger triangle as shown in Figure 2. Then glue the yellow triangles in place on the green construction paper.

© Prentice-Hall, Inc. Human Biology and Health H ■ 9

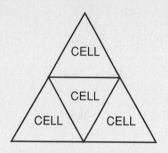

Figure 2

4. Cut out the larger yellow triangle that you glued on the green construction paper. Use the larger yellow triangle to make four identical triangles from the remainder of the green construction paper.
5. Cut out the four green triangles. Label each triangle with the word TISSUE.
6. Place the four green triangles on the sheet of blue construction paper so that they together form a larger triangle as shown in Figure 3. Then glue the green triangles in place on the blue construction paper.

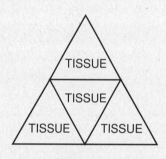

Figure 3

7. Cut out the larger green triangle that you glued on the blue construction paper. Use the larger green triangle to make four identical triangles from the remainder of the blue construction paper.
8. Cut out the four blue triangles. Label each triangle with the word ORGAN.
9. Place the four blue triangles on the sheet of red construction paper so that they together form a larger triangle as shown in Figure 4. Then glue the blue triangles in place on the red construction paper.

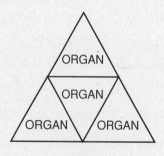

Figure 4

Name _____ Class _____ Date _____

10. Cut out the larger blue triangle that you glued on the red construction paper. Use the larger blue triangle to make four identical triangles from the remainder of the red construction paper.

11. Cut out the four red triangles. Label each triangle with the words ORGAN SYSTEM.

Observations

1. What is the color of the smallest triangle? _____

What are the words on this triangle? _____

2. What is the color of the largest triangle? _____

What are the words on this triangle? _____

3. Which triangles made up the triangle called TISSUE? _____

4. Which triangles made up the triangle called ORGAN? _____

5. Which triangles made up the triangle called ORGAN SYSTEM?

_____

Analysis and Conclusions

1. Did the triangle called ORGAN also contain cells? Explain your answer. _____

_____

_____

_____

2. Did the triangle called ORGAN SYSTEM also contain cells? Explain your answer. _____

_____

_____

_____

3. Write the levels of organization in order from largest to smallest. Start with the largest.

_____

_____

Name _____ Class _____ Date _____

Activity
The Human Body

CHAPTER 1

What Is a Tissue?

Similar cells that perform the same function are called tissues. The bodies of humans as well as most other animals contain four basic types of tissues: muscle, connective, nerve, and epithelial.

Muscle tissue has the ability to contract, or shorten. By contracting and thus pulling on bones, muscle tissue makes the body move. Connective tissue provides support for the body and connects all its parts. Nerve tissue carries messages back and forth between the brain and the spinal cord and all parts of the body. Epithelial tissue forms a covering over the outside and the inside of the body, protecting against injury and keeping out disease-causing invaders.

Examine Figures 1 through 4 and answer the questions that follow.

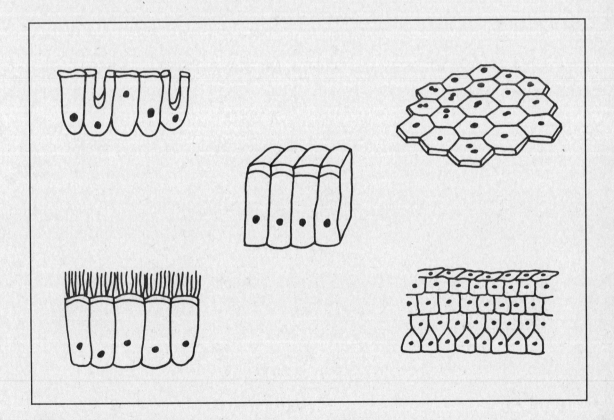

Figure 1

1. What kind of tissue does Figure 1 show? _____

2. How many kinds of this tissue are shown? _____

© Prentice-Hall, Inc. Human Biology and Health H ■ 13

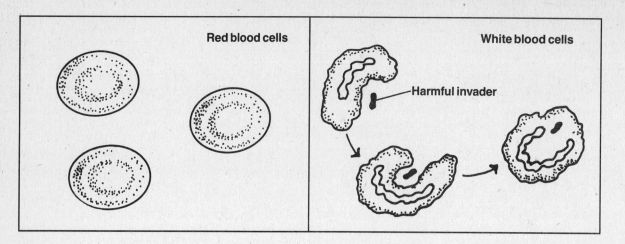

Figure 2

3. Blood is an example of which type of tissue? _____

4. Which type of cells carry oxygen to all parts of the body?
   _____

5. Which type of cells attack harmful invaders? _____

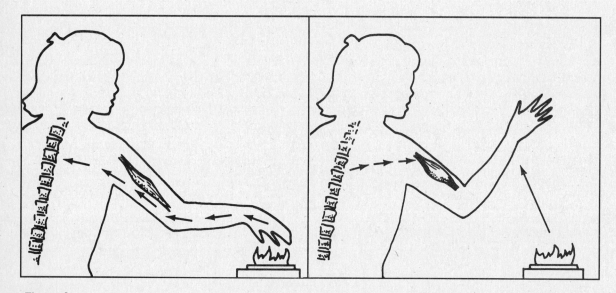

Figure 3

6. In the diagram on the left, is the nerve tissue carrying messages to or from the hand? _____

7. In the diagram on the right, is the nerve tissue carrying messages to or from the hand? _____

14 ■ H Human Biology and Health

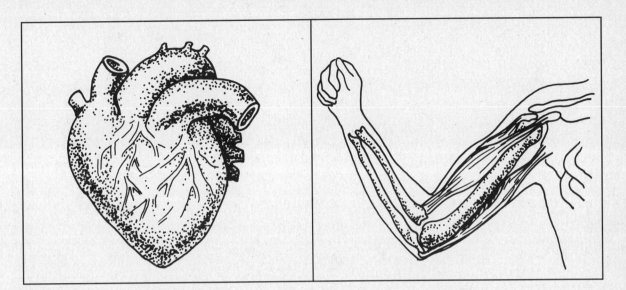

Figure 4

8. Which type of tissue makes up the heart? _____

9. What are some other parts of the body that contain this type of tissue? _____

Name _____ Class _____ Date _____

Activity
The Human Body — CHAPTER 1

Medical Terminology Made Easy

Many medical terms refer to various parts of the body. By learning to recognize the "word roots" for some of these parts, you can understand complicated medical terms.

The following is a list of word roots, many of them derived from Latin or Greek, for various structures of the body.

cardio- = heart *derm-* = skin *gastro-* = stomach
hepat- = liver *myo-* = muscle *nephro-* = kidney
neuro- = nerve *odont-* = tooth *osteo-* = bone
oto- = ear

Using the list of word roots as a guide, see if you can match each definition in Column I with the correct medical term in Column II. Knowledge of the following word endings, or suffixes, and word roots will also be helpful: *-itis* (infection, inflammation), *-ology* (the study of), *-ectomy* (surgical removal), *-osis* (abnormal or diseased), *-gram* (recorded), *electro-* (electricity), *tachy-* (swift), and *-algia* (pain).

Column I

_____ 1. Inflammation of the stomach

_____ 2. Branch of medicine concerned with the study of the nervous system

_____ 3. Malfunction of the kidneys

_____ 4. Inflammation of the liver

_____ 5. Abnormally rapid beating of the heart

_____ 6. Small mirror used by dentists to view the teeth

_____ 7. Infection of the bone marrow by bacteria

_____ 8. Pain in a muscle or muscles

_____ 9. Surgical removal of a kidney

_____ 10. Graph showing the heartbeat

_____ 11. Destruction of bone by surgery or disease

_____ 12. Disease in which pus forms at the roots of the teeth

_____ 13. Injection of fluid medicine under the skin by means of a syringe

Column II

A. Dermatologist
B. Electrocardiogram
C. Endocarditis
D. Gastrectomy
E. Gastritis
F. Hepatitis
G. Hypodermic
H. Myalgia
I. Myasthenia gravis
J. Nephrosis
K. Nephrectomy
L. Neuralgia
M. Neurology
N. Odontoscope

© Prentice-Hall, Inc. Human Biology and Health H ■ 17

_____ 14. Fungus infection of the external ear and ear canal

_____ 15. General inflammation of the peripheral nervous system

_____ 16. Inflammation of the heart lining

_____ 17. Surgical removal of a part of the stomach

_____ 18. Severe pain along a peripheral nerve

_____ 19. Disease in which the muscles are weak and tire easily

_____ 20. Specialist in diagnosis and treatment of skin disorders

O. Osteoclasia
P. Osteomyelitis
Q. Oomycosis
R. Periodontitis
S. Polyneuritis
T. Tachycardia

Name _____ Class _____ Date _____

Activity   The Human Body   CHAPTER **1**

Describing Skin

Being able to read illustrations is an important skill not only in biology but in other areas of study as well. In this activity you will interpret an illustration to identify the structures and function of the skin. The figure below shows a cross section of the skin.

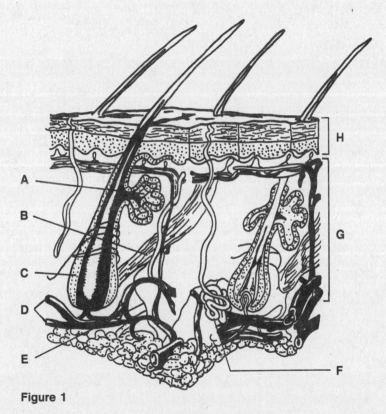

Figure 1

1. Match the letter of each structure in the illustration above with the correct name in the following list.

    _____ Blood vessels   _____ Dermis (bottom layer)

    _____ Epidermis (top layer) _____ Oil gland

    _____ Hair follicle   _____ Muscle

    _____ Nerve   _____ Sweat gland

© Prentice-Hall, Inc.   Human Biology and Health   H ■ 19

2. What is the most important function of the integumentary system? How does it perform this function? _____

_____

_____

_____

Name _____ Class _____ Date _____

Laboratory Investigation

CHAPTER 1 ■ The Human Body

Looking at Human Cheek Cells

Problem
What are the characteristics of some typical human cells?

Materials (*per pair of students*)
microscope   medicine dropper
glass slide   methylene blue
coverslip   paper towel
toothpick

Procedure 🧪 📷 👁
1. Place a drop of water in the center of the slide.
2. Using the flat end of the toothpick, gently scrape the inside of your cheek. Although you will not see them, cells will come off the inside of your cheek and stick to the toothpick.
3. Stir the scapings from the same end of the toothpick into the drop of water on the slide. Mix thoroughly and cover with the coverslip.
4. Place the slide on the stage of the microscope and focus under low power. Examine a few cells. Focus on one cell. Sketch and label the parts of the cell. (Refer to Figure 1–3 on page H17 in your textbook for the basic parts of the cell.)
5. Switch to high power. Sketch and label the cell and its parts.
6. Remove the slide from the stage of the microscope. With the medicine dropper, put one drop of methylene blue at the edge of the coverslip. **CAUTION**: *Be careful when using methylene blue because it may stain the skin and clothing*. Place a small piece of paper towel at the opposite edge of the coverslip. The stain will pass under the coverslip. Use another piece of paper towel to absorb any excess stain.
7. Place the slide on the stage of the microscope again and find an individual cell under low power. Sketch and label that cell and the cell parts that you see.
8. Switch to high power and sketch and label the cell and its parts.

Observation
How are cheek cells arranged with respect to one another? _____

_____

_____

_____

© Prentice-Hall, Inc.   Human Biology and Health   H ■ 21

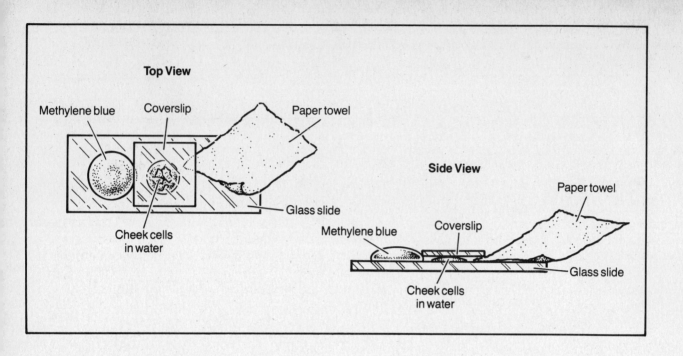

Analysis and Conclusions

1. What is the advantage of staining the cheek cells? _____

2. Explain why the shape of cheek cells is suited to their function. _____

3. Based on your observations, to which tissue type do cheek cells belong? _____

4. **On Your Own** Examine some other types of human cells, such as muscle, blood, or nerve, under the microscope. Sketch and label the parts of the cell. How does this cell compare with the cheek cell?

Answer Key

CHAPTER 1 ■ The Human Body

Chapter Discovery: Discovering Levels of Organization
Observations 1. Yellow, Cells **2.** Red, Organ system **3.** Cells **4.** Tissues
5. Organs **Analysis and Conclusions**
1. Yes. The ORGAN triangle is made of tissues and tissues are made of cells. **2.** Yes. The organ system triangle is made of organs, which are made of tissues. And tissues contain cells. **3.** Organ system; organ; tissue; cell

Activity: What Is a Tissue?
1. Epithelial **2.** 5 **3.** Connective **4.** Red blood cells **5.** White blood cells **6.** From the hand **7.** To the hand **8.** Muscle tissue **9.** Muscle tissue is also found attached to bones and lines the digestive system and blood vessels.

Discovery Activity: Medical Terminology Made Easy
1. E **2.** M **3.** J **4.** F **5.** T **6.** N **7.** P
8. H **9.** K **10.** B **11.** O **12.** R **13.** G
14. Q **15.** S **16.** C **17.** D **18.** L **19.** I
20. A

Discovery Activity: Describing Skin
1. D, H, C, E, G, A, B, F **2.** The most important function of the integumentary system is protection. It performs this function by serving as a barrier against harmful invaders and injury; helping to regulate body temperature; and providing protection from ultraviolet radiation from the sun.

Laboratory Investigation: Looking at Human Cheek Cells
Observations Cheek cells are arranged in a pattern like overlapping paving stones. Students will be looking at several layers of cells **Analysis and Conclusions 1.** Staining the cells makes it easier to see their shape and arrangement. **2.** The shape of the cells allows them to be arranged in overlapping layers to form the covering tissue of the skin. **3.** Cheek cells are epithelial tissue.
4. Answers will vary depending on the types of cells students choose.

Contents

CHAPTER 2 ■ Skeletal and Muscular Systems

Chapter Discovery

*Discovering Bones and Muscles.............................H27

Chapter Activities

*Discovery Activity: Muscle Fatigue H31
*Discovery Activity: Making a Model of the Arm H33
*Activity: Scrambled Bones H35

Laboratory Investigation Worksheet

Observing Bones and Muscles..............................H37
    (**Note:** *This investigation is found on page H46 of the student textbook.*)

Answer Key H39

*Appropriate for cooperative learning

Name _____ Class _____ Date _____

Chapter Discovery CHAPTER 2

Skeletal and Muscular Systems

Discovering Bones and Muscles

In this activity you will discover the functions of bones and muscles and their relationship to each other.

Part A Bones

1. Bend your right arm at the elbow. Then using your left hand, feel the bone that runs along the underside of your lower arm from your elbow to your wrist. How would you describe this bone? What do you think its function is?

   _____
   _____
   _____

2. Run your fingers along the top of your lower leg. Locate the bone that runs from your knee to your ankle. Describe this bone. What do you think its function is? How is this bone similar to the bone in your lower arm?

   _____
   _____
   _____
   _____

3. Using your fingers, gently tap the top of your head in several places. Then tap your forehead. How would you describe the bone in these places? What function do you think this bone has?

   _____
   _____
   _____

4. Run your fingers along the sides of your body between your waist and the area under your arms. You will find a series of bones on each side. Describe the bones. What is the function of these bones?

   _____
   _____

© Prentice-Hall, Inc. Human Biology and Health H ■ 27

5. Place your hands on a desk or table. Then move your fingers as if you were typing or playing the piano. As you watch the top of your hand, what bones do you see moving? How would you describe these bones?

___

6. Take off your shoes and socks and look closely at your feet. Wiggle your toes. How do these bones compare with the bones in your fingers?

___

7. Gently push the tip of your nose with your index finger. Then taking the tip of your nose between your thumb and index finger, gently move it back and forth. Describe the material that makes up your nose.

___

8. Gently bend the tops of your ears. Describe the material that makes up your ears.

___

9. Locate your backbone. Bending forward, run your hand up and down along the backbone. Describe your backbone. How is it different from the other bones that you have felt or observed?

___

Part B Muscles

1. At the same time you bend and straighten your arm at the elbow, place your other hand around your upper arm. See Figure 1. Where are the muscles located that cause your arm to move? Which muscle straightens the arm? Which makes it bend?

___

Name _____ Class _____ Date _____

Figure 1

2. Extend your arms straight out and drop them at your sides. Then, keeping them straight, bring your arms directly in front of you. Do this movement ten times. Where are the muscles that cause this movement located?

_____
_____

3. Raise your heels off the floor so that you are standing on tiptoe. Lower your heels back down so that your feet are flat against the floor. Do this movement ten times. Where are the muscles that are involved in this movement located?

_____
_____

4. Smile and then relax your face ten times. Where are the muscles that cause you to smile located?

_____
_____

5. Frown and then relax your face ten times. Where are the muscles that enable you to frown?

_____
_____

6. Inhale deeply, then exhale. Repeat this action ten times. Where are the muscles that allow you to take a breath located?

_____
_____

7. Carefully bend over and touch your toes. Do this action five times. Where are the muscles that allow you to do this movement located? Do you feel other muscles that are being stretched as you bend down?

_____
_____

Critical Thinking and Application

1. Based on your observations, what are three functions of bones?

_____
_____
_____
_____

2. Your nose and ears are made of cartilage. How does cartilage differ from bone?

_____
_____
_____

3. Muscles move bones. Choose two movements that you performed in Part B. Then describe the actions of the muscles and bones involved in making these movements.

_____
_____
_____
_____
_____

Name _____ Class _____ Date _____

Activity
Skeletal and Muscular Systems

CHAPTER 2

Muscle Fatigue

Anyone who has performed strenuous exercise knows that muscles can soon begin to ache, or show fatigue. In this activity you will examine how quickly some of your muscles begin to tire.

Rest your elbow on the desk so that the palm of your hand is facing toward you. Open and close your hand forcefully as many times as you can in 30 seconds. Repeat four more times, each time recording your results in the Data Table.

Stand up. Hold a book in your hand with your arm positioned straight down at your side. Keeping your arm straight, lift the book to shoulder height. Then lower it. Count the number of times you can raise and lower your arm in 30 seconds. Repeat four more times and record your results in the Data Table.

DATA TABLE

Trial	Number of Fists Made	Number of Arm Lifts
1		
2		
3		
4		
5		

1. What conclusions can you draw from your results? _____
_____
_____

You may want to obtain results for several other class members or friends. On a sheet of graph paper, graph these results using a different colored pencil to represent each individual.

© Prentice-Hall, Inc.

Human Biology and Health H ■ 31

2. Are there any differences? Might they be related to age, sex, or the physical fitness of the individual? _____

_____

_____

_____

Name _____ Class _____ Date _____

Activity

Skeletal and Muscular Systems

CHAPTER **2**

Making a Model of the Arm

To do this activity you will need a large sheet of posterboard, scissors, a metric ruler, a pencil, an 18-cm-long rubber band (cut into 2 strands lengthwise), a paper fastener, and a stapler.

Procedure

1. On the posterboard, draw outlines of the arm as shown in Figure 1. Make piece A and piece B about 10 cm wide and 25 cm long. Then cut out pieces A and B.

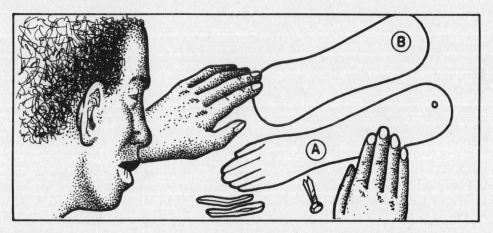

Figure 1

2. Place piece A over piece B as shown in Figure 2.
3. Put the paper fastener through both pieces at point C. Spread the fastener to close it.

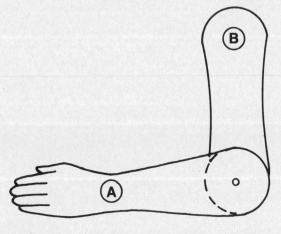
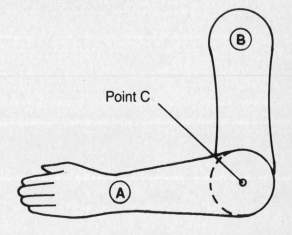

Figure 2

© Prentice-Hall, Inc.

Human Biology and Health H ■ 33

4. Staple one end of the rubber band strand at point D and the other end at point E.
5. Staple the end of the second rubber band strand at point F and the other end at point G.

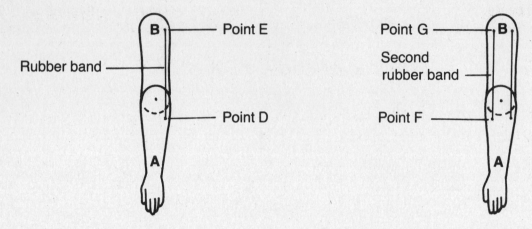

Figure 3

6. Pull the second rubber band upward as shown in Figure 4.

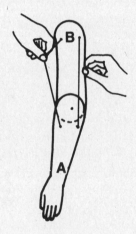

Figure 4

Observations and Conclusions

1. When you pull the rubber band, what happens? _____
   _____

2. What kind of joint is the human elbow? _____

3. How can you get your model arm to straighten again? _____
   _____

4. To raise your own forearm at the elbow, what must contract? _____
   What must relax? _____

Name _____ Class _____ Date _____

Activity

Skeletal and Muscular Systems

CHAPTER 2

Scrambled Bones

Listed below are the names of various bones of the human body, but the names are written in code! Decode these names. Then locate each bone on the skeleton. Write the number of the decoded bone on the correct blank next to the skeleton. KEY TO CODE: The correct letter is the letter that directly comes before the given letter in the alphabet. Example: B = A, S = R

1. TUFSOVN _____
2. QBUFMMB _____
3. VMOB _____
4. DSBOJVN _____
5. SJC _____
6. UJCJB _____
7. WFSUFCSB _____
8. IVNFSVT _____
9. DMBWJDMF _____
10. GJCVMB _____
11. TDBQVMB _____
12. GFNVS _____
13. SBEJVT _____
14. UBSTBMT _____
15. DBSQBMT _____

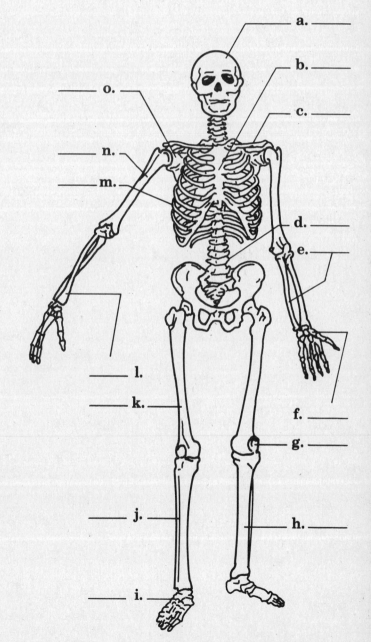

© Prentice-Hall, Inc.

Human Biology and Health H ■ 35

Name _____ Class _____ Date _____

Laboratory Investigation

CHAPTER 2 ■ Skeletal and Muscular Systems

Observing Bones and Muscles

Problem
What are the characteristics of bones and muscles?

Materials *(per group)*
2 chicken leg bones
tiny piece of raw, lean beef
2 dissecting needles
methylene blue
2 jars with lids
knife
vinegar
medicine dropper
water
2 glass slides
coverslip
microscope
paper towel

Procedure
Part A
1. Place one chicken leg bone in each of the two jars.
2. Fill one of the jars about two-thirds full with vinegar. Cover both jars.
3. After five days, remove the bones from the jars. Rinse each bone with water.
4. With a knife, carefully cut each of the bones in half. **CAUTION:** *Be careful when using a knife.* Examine the inside of each bone.

Part B
1. Place the tiny piece of raw beef on one of the glass slides. With a medicine dropper, place a drop of water on top of the beef.
2. With the dissecting needles, carefully separate, or tease apart, the fibers of the beef. **CAUTION:** *Be careful when using dissecting needles.*

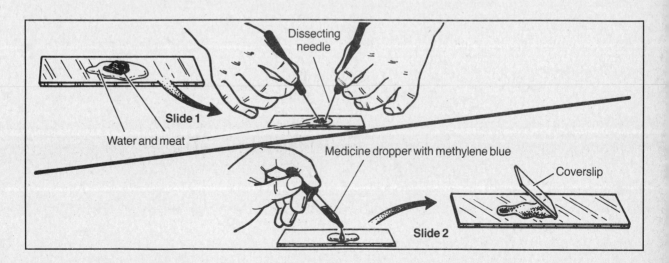

© Prentice-Hall, Inc.   Human Biology and Health H ■ 37

3. Transfer a few fibers to the second slide. Add a drop of methylene blue. **CAUTION:** *Be careful when using methylene blue because it may stain the skin and clothing.* Cover with a coverslip. Use the paper towel to absorb any excess stain.
4. Examine the slide under the microscope.

Observation
1. How do the two bones differ in texture and flexibility? Describe the appearance of the inside of each bone. _____

2. Describe the appearance of the beef under a microscope. _____

Analysis and Conclusions
1. What has happened to the minerals and the marrow within the bone that was put in vinegar? How do you know this? _____

2. Why was one bone put in an empty jar? _____

3. What type of muscle tissue did you observe under the microscope? _____

4. How does the structure of the muscle tissue aid in its function? _____

5. **On Your Own** Repeat this investigation using substances other than vinegar.

Answer Key

CHAPTER 2 ■ Skeletal and Muscular Systems

Chapter Discovery: Discovering Bones and Muscles

Part A Bones 1. Accept all reasonable answers. **2.** Accept all reasonable answers. Students should recognize that the lower leg and lower arm bones are both long and smooth and that they are important for movement. **3.** The bone is hard, smooth, and continuous over a large area. Its function is to provide shape and support for the head and to protect the brain. **4.** The bones that form the ribs are in series. The ribs provide support, give shape to the body, and protect the heart and the lungs. **5.** Bones of the fingers and wrist move. These bones are small and thin. **6.** The bones of the toes are small and thin and are very similar to the bones in the fingers. **7.** Accept all reasonable answers. **8.** Accept all reasonable answers. **9.** Backbone feels bumpy and seems to be in segments. This is different from other bones, which are smooth and continuous. Backbone also is able to bend, yet feels hard and inflexible. **Part B Muscles 1.** The muscles that bend and straighten the arm are located at the front and back of the upper arm. The muscle in the back of the upper arm straightens the arm, while the muscle in the front of the upper arm bends it. **2.** The muscles are located across the chest. **3.** The muscles are located in the backs of the calves and in the feet. **4.** The muscles are located at the corners of the mouth and in the cheeks. **5.** The muscles are located at the corners of the mouth and in the forehead between the eyebrows. **6.** The muscles are located in the chest. **7.** The muscles are located in the lower back as well as in the backs of legs. **Critical Thinking and Application 1.** Answers may include provide for movement, give shape and support, and protect organs. **2.** Cartilage is flexible and soft, while bone is inflexible and hard. **3.** Answers will vary.

Discovery Activity: Muscle Fatigue

Answers in the Data Table will vary, but the numbers should decrease with each trial.
1. The muscles of the fingers, hand, and arm get tired with continuous use. Fewer tasks can be performed in the later trials. **2.** Elderly and very young people and those that are not very physically fit will show a very definite decrease with each trial.

Discovery Activity: Making a Model of the Arm

Observations and Conclusions 1. The arm, from the elbow to the hand, moves upward. **2.** Hinge **3.** Pull on the other rubber band **4.** Biceps. Triceps

Activity: Scrambled Bones

1. Sternum **2.** Patella **3.** Ulna **4.** Cranium **5.** Rib **6.** Tibia **7.** Vertebra **8.** Humerus **9.** Clavicle **10.** Fibula **11.** Scapula **12.** Femur **13.** Radius **14.** Tarsals **15.** Carpals **a.** 4 **b.** 1 **c.** 9 **d.** 7 **e.** 3 **f.** 15 **g.** 2 **h.** 6 **i.** 14 **j.** 10 **k.** 12 **l.** 13 **m.** 5 **n.** 8 **o.** 11

Laboratory Investigation: Observing Bones and Muscles

Observations 1. The bone submerged in vinegar is more flexible and not as hard and solid as the bone in the empty jar. In comparison to the control, the bone submerged in vinegar has lost some of its hard, minerallike texture. **2.** The beef should resemble long, fiberlike threads under the microscope. **Analysis and Conclusions 1.** Some of the minerals and marrow dissolved in the vinegar. This can be determined by comparing the bone submerged in vinegar to the control. **2.** This bone acts as a control. **3.** Skeletal, or striated, muscle. **4.** The fiberlike threads permit movement (contraction and relaxation) to occur. **5.** Answers will vary, depending on the substances students choose. Possible substances are vegetable oil, weak hydrochloric acid, and milk.

© Prentice-Hall, Inc.

Contents

CHAPTER 3 ■ Digestive System

Chapter Discovery
*How Much Fat? H43

Chapter Activities
*Discovery Activity: Classifying Foods H47
*Problem-Solving Activity: Categories of Digestion H49
*Discovery Activity: Nutrition Survey H51
*Discovery Activity: Snack Survey H53
Discovery Activity: What Color Is Your Cracker? H55
*Problem-Solving Activity: The Structure of Teeth H57
Problem-Solving Activity: Calculating Your Nutritional Budget H59
Discovery Activity: Identifying Nutrients H61
*Problem-Solving Activity: Digestive Trivia H63

Laboratory Investigation Worksheet
Measuring Calories Used H65
    (**Note:** *This investigation is found on page H74 of the student textbook.*)

Answer Key H67

*Appropriate for cooperative learning

Name _____ Class _____ Date _____

Chapter Discovery — Digestive System — CHAPTER 3

How Much Fat?

Problem
How can the fat content of ground beef be determined?

Materials
3 samples of ground beef:
  regular
  lean
  leanest (or extra lean)
3 800-mL beakers
heat source
beaker tongs or heat-resistant gloves
100-mL graduated cylinder
stirring rod
spoon
glass-marking pencil
balance scale

Procedure

1. Examine the labels of the three ground beef samples. Record the information from each label in Data Table 1.
2. Label the three beakers 1, 2, and 3, with the glass-marking pencil.
3. Use the balance scale to obtain 100 g of ground beef sample 1. Place the ground beef in beaker 1.
4. Fill the beaker about three-fourths full with water.
5. Place the beaker on the heat source. **CAUTION:** *Always observe safety rules when using a heat source.* Stir the contents of the beaker thoroughly and bring to a boil.
6. Use the tongs or heat-resistant gloves to remove the beaker from the heat. Allow the meat to settle to the bottom of the beaker.
7. The fat will form a layer above the water. Carefully pour off the fat layer into the graduated cylinder. If some of the fat particles are difficult to remove, wait until the mixture cools. Then scoop up any solid fat particles and add them to the fat in the graduated cylinder.
8. Find the volume of fat by reading its level in the graduated cylinder. Record your measurement in Data Table 2.
9. You may assume that 1 mL of fat is about equal to 1 g of fat. Record the mass of the fat in Data Table 2.

© Prentice-Hall, Inc.   Human Biology and Health

10. Because your original sample was 100 g, you can find the percentage of fat easily by dividing the mass of the fat by 100 g. Then multiply this number by 100. Record the percentage of fat in Data Table 2.
11. Repeat steps 3 through 10 for beef samples 2 and 3.

Observations

Data Table 1

Sample	Stated Percentage of Fat	Cost Per 100 g
1		
2		
3		

Data Table 2

Sample	Mass (g)	Volume Fat (mL)	Mass Fat (g)	Percentage of Fat
1				
2				
3				

Analysis and Conclusions

1. Which sample of ground beef had the lowest percentage of fat? Which had the highest?

   _____
   _____

2. How do the experimental values for the percentages of fat compare with the percentages of fat stated on the label of each sample? What might account for the differences between the two values?

   _____
   _____
   _____
   _____
   _____
   _____

Name _____ Class _____ Date _____

3. What appears to be the relationship between cost and percentage of fat for ground beef? What does this tell you about the kind of meat people want to buy?

_____
_____
_____
_____

4. Which of the three samples of ground beef would you be most likely to buy? Explain your answer.

_____
_____
_____
_____

Name _____ Class _____ Date _____

Activity

Digestive System

CHAPTER 3

Classifying Foods

Study the drawings of different foods. Then follow the directions.

a. Circle the foods that are high in protein.
b. Draw a square around the foods that are high in carbohydrates.
c. Write an X over the foods that have a high fat content.
d. Write a C over the foods that have a high vitamin C content.
e. Draw a triangle around the foods that are high in vitamin A.
f. Draw a line under the foods that belong to the vegetable and fruit group.
g. Draw a line above the foods that belong to the milk group.
h. Write an I over the foods that are high in iodine.

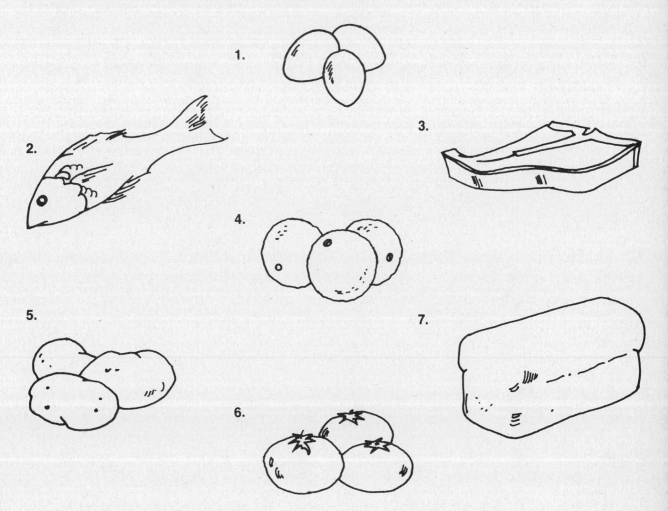

© Prentice-Hall, Inc.

Human Biology and Health H ■ 47

8.

9.

10.

11.

12.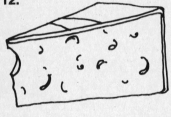

13.

14.

48 ■ H Human Biology and Health

Activity

Digestive System — CHAPTER 3

Categories of Digestion

Four categories related to digestion are listed at the top of the chart. Can you think of items for each category that begin with the letters shown at the left? You can list more than one item, but try to think of at least one for each category.

	Structure Involved in Digestion	Food	Digestive Secretion or Enzyme	Vitamin or Mineral
M				
I				
L			lipase lactase	✕
S				
T				
P	pancreas pharynx			

Human Biology and Health

Name _____ Class _____ Date _____

Activity
Digestive System

CHAPTER 3

Nutrition Survey

It is important that the foods you eat come from the four essential food groups. Foods in the food groups contain carbohydrates, proteins, fats, and vitamins. During the next five days keep a record of all the foods you eat during breakfast, lunch, and dinner. Be sure to include snacks, soft drinks, and so on. Construct a chart listing the carbohydrates, proteins, fats, and vitamins that the food contains. You may have to use a reference book to find out what vitamins some foods contain if the information is not listed on the package label.

Food	Carbohydrates	Proteins	Fats	Vitamins

© Prentice-Hall, Inc.

Human Biology and Health   H ■ 51

Food	Carbohydrates	Proteins	Fats	Vitamins

1. What type of nutrient do you eat most? _____

2. What vitamins are lacking in your diet? _____

3. What should you do to make sure your diet is well balanced? _____

Name _____ Class _____ Date _____

Activity

Digestive System

CHAPTER 3

Snack Survey

What is your favorite snack? Ask at least 20 of your classmates to choose their three favorite snacks from the list below. Then ask 20 adults to select their three favorite snacks. Determine the three most popular snacks for each group. Record the results of these two surveys on a separate sheet of paper. Construct a graph showing your results on a sheet of graph paper.

Food	Approximate Calories	Food	Approximate Calories
Candy bar per oz.	150	Chocolate chip cookie (1)	50
Cola, 12 oz.	145	Cheese pizza, 1 slice	185
Potato chips, (10)	115	Pretzel twist (1)	25
Ice cream	140	Brownie	150
Orange juice, 1 cup	110	Fruit yogurt	270
Hot dog with bun	225	Hamburger with bun, ¼ lb.	420
Milkshake	330	French fries, regular serving	215
Beef taco	200	Apple or orange	70
Banana	85	Milk, 1 glass	160
Doughnut	180	Peanut butter sandwich	325
Bean burrito	320	Carrot sticks (4)	15

1. Which snacks were the most popular for each group? _____

2. If there were differences in favorites between the two groups, what might be some reasons for these differences? _____

3. What is the average number of Calories for the three foods selected by each group? (To find this number, add the Calories for each of the three snacks for all the people in each group and then divide by 3.) _____

4. Is there a difference in the number of Calories contained in each group for their favorite snacks? _____

© Prentice-Hall, Inc.

Human Biology and Health

Name _____ Class _____ Date _____

Activity
Digestive System

CHAPTER 3

What Color Is Your Cracker?

Benedict's solution is a blue solution that turns greenish yellow when a little sugar is present and turns red or orange when a lot of sugar is present.

1. Obtain two test tubes, a cracker, and some Benedict's solution.
2. Use a straw to fill one test tube about one-fourth full of saliva and leave the other empty.
3. Put a small piece of cracker into each test tube and allow them to stand overnight.
4. On the following day, place a few drops of Benedict's solution into each test tube.
5. Place the test tubes in a beaker that is half full of water.
6. Carefully heat the beaker of water with a Bunsen burner. **CAUTION:** *Be very careful when using a Bunsen burner.*
7. Record your observations in the chart below.

Test Tube With Saliva	Test Tube Without Saliva

What conclusion can you draw from this experiment?

_____
_____
_____
_____
_____
_____
_____
_____
_____
_____
_____

© Prentice-Hall, Inc.

Human Biology and Health

Name _____ Class _____ Date _____

Activity
Digestive System

CHAPTER 3

The Structure of Teeth

Part A

Study the four types of teeth in the drawing of the upper jaw and fill in the chart below.

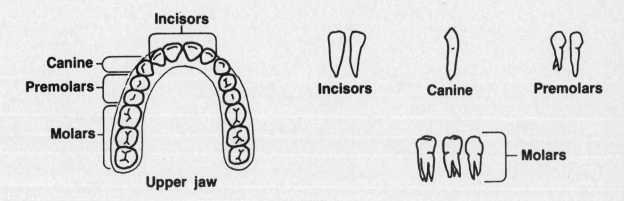

Type of Tooth	Number of Teeth in Upper Jaw
Molar	
Premolar	
Canine	
Incisor	
Total	

Part B

Using a flashlight and a mirror, examine your own teeth. Locate your incisors, canines, premolars, and molars. In the drawing of the upper and lower jaws on page 58, map your teeth onto the diagrams. Place an "O" on the drawing of any teeth you have. Place an "X" on the drawing of any teeth that you do not have. Label the four types of teeth.

© Prentice-Hall, Inc.

Human Biology and Health H ■ 57

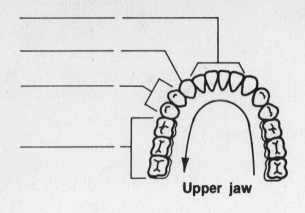
Upper jaw

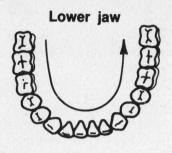
Lower jaw

Part C

A full set of adult teeth is expressed with the following dental formula:

Molars	Premolars	Canines	Incisors	Molars	Premolars	Canines	Incisors
$\frac{3}{3}$	$\frac{2}{2}$	$\frac{1}{1}$	$\frac{2}{2}$	$\frac{2}{2}$	$\frac{1}{1}$	$\frac{2}{2}$	$\frac{3}{3}$

The numbers on the top lines represent the numbers of teeth in the upper jaw, while the numbers below the lines represent the teeth in the lower jaw. Notice the center line that divides the numbers into left and right sides. The numbers to the left designate the teeth in the right side of the mouth, while the numbers to the right of the center line represent the teeth in the left side of the mouth. The teeth are also divided into the four types as they appear from the back right, around the front, to the back left of the mouth. See arrows on the diagrams.

In the space below, write the dental formula for your mouth.

Name _____ Class _____ Date _____

Activity
Digestive System

CHAPTER 3

Calculating Your Nutritional Budget

To find out the nutritional value of the foods you eat, read the labels on boxes and cans. Then follow these directions.

Obtain five different cans or boxes of food. To calculate the percentage of sugar in a serving, look at the label and find out how many grams of sugar there are in one serving of that food. Multiply this number by 4 Calories (1 gram of sugar provides 4 Calories). This tells you the Calories' worth of sugar per serving. Now check the label for the number of Calories in a serving. Divide the number of Calories' worth of sugar per serving by the label's number of Calories in a serving. Then multiply this number by 100 and you will get the percentage of sugar in a serving. Record your results in the chart.

Food	Percentage of Sugar Calories

1. If the total daily recommended intake of carbohydrate Calories is 58 percent, which foods on your chart exceed this amount? _____

Using the same cans or boxes of food, calculate the percentage of fat Calories in a serving of each. To do so follow the same calculations as in finding the sugar percentage, but instead of multiplying the number of grams by 4, multiply it by 9 (1 gram of fat provides 9 Calories). Record your results in the chart.

Food	Percentage of Fat Calories

2. If the total daily recommended intake of fat Calories is 30 percent, which foods on your chart exceed this amount? _____

© Prentice-Hall, Inc.

Name _____ Class _____ Date _____

Activity
Digestive System

CHAPTER 3

Identifying Nutrients

Carbohydrates (starches and sugars), fats, proteins, minerals, water, and vitamins are nutrients. Some of these nutrients can be detected by taste, while others cannot. Therefore, scientists use certain tests to identify the presence of nutrients. In this activity you will learn to test for proteins and fats in gelatin, a piece of raw potato, corn, bean seeds, and cooking oil.

Testing for Proteins

Using a medicine dropper, fill a test tube about one-third full of gelatin solution. To this, add 10 drops of Biuret solution. **CAUTION:** *Biuret solution will burn skin and clothing.* Hold the test tube against a sheet of white paper. If the mixture has turned a violet color, then protein is present. Repeat this procedure using the four other foods. Record your results in the chart. Use a plus sign (+) to indicate the presence of a protein and a negative sign (−) to indicate its absence.

Testing for Fats

Place a few drops of cooking oil on a piece of brown paper. Hold the paper up to the light and look through the spot. If the spot is greasy and translucent, a fat is present. Repeat this procedure for the four other foods. Record your results in the chart. Use a plus sign (+) to indicate the presence of a fat and a negative sign (−) to indicate its absence.

Food	Fat	Protein
Gelatin		
Potato		
Corn		
Bean		
Cooking oil		

© Prentice-Hall, Inc.

Human Biology and Health

Name _____ Class _____ Date _____

Activity

Digestive System

CHAPTER 3

Digestive Trivia

How many of these "trivia" questions on digestion can you answer?

1. How long is the average adult's esophagus? _____

2. In the early 1800s, Dr. William Beaumont gained important information about human digestion from studying a man who, due to a gunshot wound, had a permanent hole in his stomach. Name the patient. _____

3. Which of the following cannot be digested by humans—sucrose, maltose, cellulose, fructose? _____

4. Name the wavelike muscle contractions that move food through the digestive system. _____

5. Humans have four types of teeth. Name them. _____

6. What is the average diameter of the large intestine? _____

7. List the three parts, or sections, of the small intestine. _____

8. What structure keeps food from "going down the wrong pipe"? _____

9. How many lobes does the human liver have? _____

10. What is the function of the appendix in humans? _____

11. How long does food stay in the stomach before entering the small intestine? _____

12. The shape of the large intestine is like an upside-down letter of the alphabet. What letter is this? _____

13. When the mucous lining and part of the stomach wall are destroyed or "eaten away," what medical problem is the result? _____

14. Name a structure of the human body that does not contain blood vessels. _____

15. Name an enzyme found in pancreatic juice. _____

16. What kind of acid is secreted by the stomach? _____

17. How long does the entire digestive process take? _____

18. What is the largest organ found inside the human body? _____

19. What is the name of the partly digested, souplike food mass that leaves the stomach? _____

20. How fast does food move through the small intestine? _____

21. What percentage of your body mass is water? _____

22. Name the largest gland in the human body. _____

23. What is the disease called in which people can literally diet themselves to death? _____

24. Approximately how many villi are found in the small intestine? _____

25. The large intestine is actually shorter than the small intestine. How did it get its name? _____

Name _____ Class _____ Date _____

Laboratory Investigation

CHAPTER 3 ■ Digestive System

Measuring Calories Used

Problem
How many Calories do you use in 24 hours?

Materials *(per student)*
pencil and paper   scale

Procedure
1. Look over the chart of Calorie rates. It shows how various activities are related to the rates at which you burn Calories. The Calorie rate shown for each activity is actually the number of Calories used per hour for each kilogram of your body mass.

Activity	Average Calorie Rate
Sleeping	1.1
Awake but at rest (sitting, reading, or eating)	1.5
Very slight exercise (bathing, dressing)	3.1
Slight exercise (walking quickly)	4.4
Strenuous exercise (dancing)	7.5
Very strenuous exercise (running, swimming rapidly)	10.5

2. Using a scale, note your weight in pounds. Convert your weight into kilograms (2.2 lb = 1 kg). Record this number.

3. Classify all your activities for a given 24-hour period. Record the kind of activity, the Calorie rate, and the number of hours you were involved in that activity.

4. For each of your activities, multiply your weight by the Calorie rate shown in the chart. Then multiply the resulting number by the number of hours or fractions of hours you were involved in that activity. The result is the number of Calories you burned during that period of time. For example, if your weight is 50 kilograms and your exercised strenuously, perhaps by running, for half an hour, the Calories you burned during that activity would be equal to $50 \times 10.5 \times 0.5 = 262.5$ Calories.

5. Add together all the Calories you burned in the entire 24-hour period.

Observations
How many Calories did you use in the 24-hour period? _____

Average Caloric Needs Chart		
	Age	Calories
Males	9–12 12–15	2400 3000
Females	9–12 12–15	2200 2500

Analysis and Conclusions

1. Explain why the values for the Calorie rates of various activites are approximate rather than exact. _____
_____
_____

2. What factors could affect the number of Calories a person used during exercise?
_____
_____
_____

3. Why do young people need to consume more Calories than adults? _____
_____
_____
_____

4. **On Your Own** Determine the number of Calories you use in a week. In a month.

66 ■ H   Human Biology and Health

Answer Key

CHAPTER 3 ■ Digestive System

Chapter Discovery: How Much Fat?
Observations Answers will vary. **Analysis and Conclusions 1.** Answers will vary depending on how students labeled their samples. **2.** Answers will vary. Differences between the two values may be due to problems in the experimental procedure, such as difficulty in recovering all the fat. Differences can also be caused by the fact that 1 mL of fat is not exactly equal to 1 g of fat, although it is quite close. Another reason for a difference between the two values could be due to the improper labeling on the package. **3.** The leanest sample of beef will probably cost the most, whereas the fattiest sample will cost the least. The leanest sample of beef is not only the most desirable but it is also the most expensive. **4.** Accept all logical answers.

Discovery Activity: Classifying Foods
Students should classify each drawing of food as follows: **1.** Protein. Vitamin A **2.** Protein. Iodine **3.** Protein. Fat **4.** Carbohydrate. Vitamin C. Fruit and vegetable **5.** Carbohydrate. Fruit and vegetable **6.** Vitamin C. Vitamin A. Fruit and vegetable **7.** Carbohydrate **8.** Carbohydrate. Fat **9.** Carbohydrate. Vitamin A. Fruit and vegetable **10.** Vitamin C. Vitamin A. Fruit and vegetable **11.** Protein. Milk **12.** Protein. Fat. Milk **13.** Fat. Vitamin A. Milk **14.** Carbohydrate. Vitamin A. Fruit and vegetable

Problem-Solving Activity: Categories of Digestion
From left to right M: mouth; macaroni, mushrooms, milk; mucus; magnesium **I:** intestines; ice cream, iced tea; intestinal juice; ion, iodine **L:** liver, large intestine; lettuce, lobster, liver **S:** stomach, small intestine, salivary glands; spaghetti, steak, sugar; saliva, sucrase; sodium **T:** teeth, tongue; taco, turkey, tomato; trypsin, thiamin, tocopherol (Vitamin E) **P:** potatoes, peach pie; pepsin, peptidase, pancreatic fluid; phosphorus, pantothenic acid, pyridoxine (Vitamin B_6)

Discovery Activity: Nutrition Survey
Answers will vary. Check charts to be sure that students have classified the nutrients correctly. You may want to supply some general nutritional information. There are many reliable sources that provide excellent material at no cost or for a small charge. These materials include menu plans and recipes. Following is a list of sources: American Heart Association, American Dietetic Association, American Diabetic Association, Local Health Departments (Bureau of Nutrition).

Discovery Activity: Snack Survey
1. Answers will vary. **2.** Answers will vary. However, students may tend to eat more fast foods, which contain more Calories, than adults do. **3.** Answers will vary. **4.** On the average, the student group will probably have more Calories than the adult group.

Discovery Activity: What Color Is Your Cracker?
Students should observe that the test tube with saliva in it turned red or orange, a positive test for sugar. From this observation, they should conclude that something in the saliva changed the starch in the cracker to sugar. From the text, they should know that this "something" is the enzyme ptyalin.

Problem-Solving Activity: The Structure of Teeth
Part A Molar 6 **Premolar** 4 **Canine** 2 **Incisor** 4 **Total** 16 **Part B Top to bottom** Incisors; Canine; Premolars; Molars **Part C** Dental formulas will vary.

Problem-Solving Activity: Calculating Your Nutritional Budget

Answers for this activity will vary. Check charts to be certain that students have calculated correctly. Check to make sure that answers are reasonable.

Discovery Activity: Identifying Nutrients

Gelatin negative for fat; positive for protein
Potato negative for fat; positive for protein
Corn positive for fat; positive for protein
Bean positive for fat; positive for protein
Cooking oil positive for fat; negative for protein

Problem-Solving Activity: Digestive Trivia

1. About 30 cm 2. Alexis St. Martin 3. Cellulose 4. Peristalsis 5. Incisors, canines, premolars, molars 6. About 6.5 cm 7. Duodenum, jejunum, ileum 8. Epiglottis 9. 4 10. May function in immunity 11. 2–6 hours 12. U 13. Ulcer 14. Hair, nails, epidermis, cartilage, or cornea 15. Trypsin, amylase, or lipase 16. Hydrochloric acid 17. 12–24 hours 18. Liver 19. Chyme 20. 7–8 cm/min 21. 50–70 percent 22. Pancreas 23. Anorexia nervosa 24. 5 million 25. From its wide diameter

Laboratory Investigation: Measuring Calories Used

Observations Answers will vary considerably, and students should be cautioned that each person burns a different number of Calories on any given day. The majority of students will burn between 1400 and 4000 Calories per day. **Analysis and Conclusions**
1. Each person's body chemistry is slightly different, so no exact rate per activity can be determined. 2. Answers may include the size of the person, how hard the person exercises, the temperature of the environment, and so on. 3. Young people are generally more active and need to consume more Calories, not only for basic metabolic functions, but also for growth. 4. Answers will vary, depending on the student.

Contents

CHAPTER 4 ■ Circulatory System

Chapter Discovery
*Simulating Blood Transfusions H71

Chapter Activities
*Discovery Activity: The Heartbeat. H75
*Problem-Solving Activity: Circulation Calculations H77
*Problem-Solving Activity: Circulation Comparisons H79
*Discovery Activity: Surveying Risk Factors Related to Heart
 Disease H83

Laboratory Investigation Worksheet
Measuring Your Pulse Rate H85
  (**Note:** *This investigation is found on page H102 of the student textbook.*)

Answer Key H87

*Appropriate for cooperative learning

Name _____ Class _____ Date _____

Chapter Discovery | Circulatory System — CHAPTER 4

Simulating Blood Transfusions

Background Information

Perhaps you or someone you know has had a blood transfusion. Blood transfusions are necessary when a person loses a great deal of blood as a result of an injury or an illness. Before a blood transfusion can be given, tests must be made to make sure that the blood groups of the donor (person giving blood) and the recipient (person receiving blood) are compatible.

What are blood groups? Human blood is classified into four basic groups—A, B, AB, and O—depending on the presence or absence of certain proteins. In this activity you will discover which blood groups can be safely mixed in a blood transfusion.

Materials

8 paper cups
medicine dropper
yellow and blue food coloring
marking pen

Procedure

1. Label four paper cups A, B, AB, and O. Below each of these letters, place the letter D for donor. These will be the blood donor cups.
2. Fill each cup half full with water. Set cup O aside for now.
3. Add 2 drops of yellow food coloring to cup A and 2 drops of blue food coloring to cup B.
4. To cup AB, add 2 drops of yellow food coloring and 2 drops of blue food coloring.
5. Label the remaining four paper cups A, B, AB, and O. Below each of these letters, place the letter R for recipient. These cups will represent the recipients' blood groups A, B, AB, and O.
6. Repeat steps 2 through 4 using these cups.
7. Using the medicine dropper, place 2 drops of group A "blood" from the donor cup into each recipient's cup. See Figure 1. If the recipient's blood does not change color, the transfusion is said to be safe. If a color change does occur, the transfusion is unsafe. Record your results for mixing group A blood with each of the recipients' blood groups in the Data Table. Write "yes" if the transfusion is safe; write "no" if it is unsafe.

© Prentice-Hall, Inc. Human Biology and Health H ■ 71

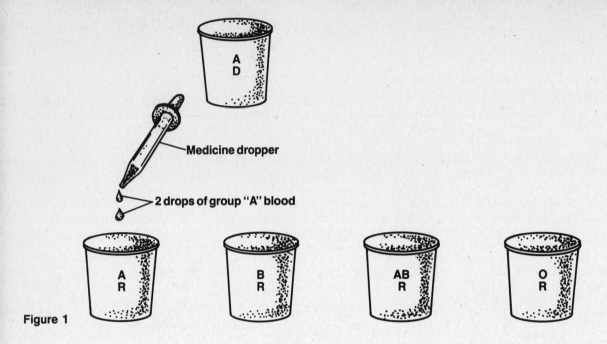

Figure 1

8. Rinse out the dropper and the recipients' cups. To fill the recipients' cups, repeat steps 2 through 4.
9. Using the dropper, place 2 drops of group B blood from the donor cup into each recipient's cup. Record your results in the Data Table.
10. Repeat steps 8 and 9, placing 2 drops of group AB blood into each recipient's cup.
11. Then repeat steps 8 and 9, placing 2 drops of group O blood into each recipient's cup.

Observations
Data Table

Donor Blood Group	Recipient Blood Group			
	A	B	AB	O
A				
B				
AB				
O				

Name _____ Class _____ Date _____

Analysis and Conclusions

1. From which blood groups can a person with group A blood safely receive?
   _____

2. Which blood groups can safely receive group B blood?
   _____

3. For which blood group would it be easiest to find a donor? Explain your answer.
   _____
   _____
   _____
   _____

4. For which blood group are donors most in demand? Explain.
   _____
   _____
   _____

5. Can a group A recipient safely receive group O blood?
   _____

Name _____ Class _____ Date _____

Activity
Circulatory System

CHAPTER 4

The Heartbeat

Purpose
In this activity you will construct a stethoscope and use it to listen to another student's heartbeat.

Materials *(per group)*
30-cm length of rubber tubing
2 small funnels
adhesive tape

Procedure
1. Attach a small funnel to each end of a length of rubber tubing. Using the adhesive tape, seal the funnel to the tubing. You have just constructed a simple stethoscope.
2. Have the student whose heartbeat you will listen to hold one funnel on his or her chest, directly over the heart.
3. With one hand, hold the second funnel to your ear. Cover your other ear with your other hand and listen carefully for a minute or so.

Observations and Conclusions
1. What types of sounds did you hear? _____

_____

_____

_____

2. What causes these sounds? _____

_____

_____

_____

3. Why do you think a doctor uses a stethoscope to listen to a person's heartbeat?

_____

_____

_____

Name _____ Class _____ Date _____

Activity
Circulatory System
CHAPTER 4

Circulation Calculations

The following problems involve some interesting facts about the blood, heart, and circulatory system. Place your answers on the lines provided.

1. If blood is 7 percent of your body mass, what is the mass of your blood in kilograms?

   Answer _____

2. If the average teenager's body contains 4.5 liters of blood, how many liters of blood would there be in a class of 30 students?

   Answer _____

3. At rest, the human heart pumps about 5 liters of blood per minute. How many liters are pumped in 1 hour? In 5 hours? In 1 day?

   Answers _____ _____ _____

4. If your heart beats 70 times per minute, how many times does it beat in 1 hour? In 1 day? In 1 year?

   Answers _____ _____ _____

5. On the average, males have about 5,200,000 red blood cells per cubic millimeter of blood, while females have only 4,200,000 red blood cells per cubic millimeter. How many more red blood cells are in 20 cubic millimeters of Kevin's blood than in Kara's?

   Answer _____

© Prentice-Hall, Inc.

Human Biology and Health H ■ 77

6. A one-cubic-millimeter sample of blood shows 8000 white blood cells and 5,200,000 red blood cells. Do you think this sample is from a male or a female? _____
How many times more red blood cells than white blood cells are there in this sample?

Answer _____

7. Red blood cells are constantly wearing out and being replaced with new ones from the bone marrow at a rate of 5 million per second. At this fantastic rate, how many red blood cells are replaced each hour?

Answer _____

8. About 45 percent of the people in the United States have group O blood. If there is one person with group AB for every 10 persons with group O, what percentage of the population has group AB blood?

Answer _____

9. Assume that 40 percent of the people in your school have group A blood. How many people would this be?

Answer _____

10. Blood is pumped out of the heart and into the aorta at the rate of 50 miles per hour. How fast is this in kilometers?

Answer _____

Name _____ Class _____ Date _____

Activity
Circulatory System

CHAPTER **4**

Circulation Comparisons

Examine each pair of drawings below. Match each drawing with the term on the right that correctly identifies it. Then tell how the two objects are similar and how they are different.

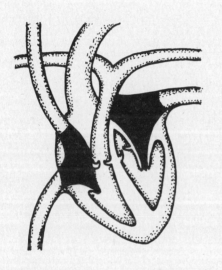 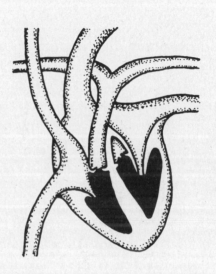

**Atria
or
Ventricles**

1. _____   _____

Similarities _____

_____

Differences _____

_____

 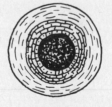

**Artery
or
Vein**

2. _____   _____

Similarities _____

_____

Differences _____

_____

© Prentice-Hall, Inc.

Human Biology and Health H ■ 79

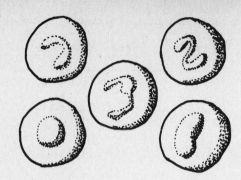

**Red Blood Cells
or
White Blood Cells**

3. _____   _____

Similarities _____
_____

Differences _____
_____

**Group A Blood
or
Group B Blood**

4. _____   _____

Similarities _____
_____

Differences _____
_____

Name _____ Class _____ Date _____

Plasma or Whole Blood

5. _____   _____

Similarities _____
_____

Differences _____
_____

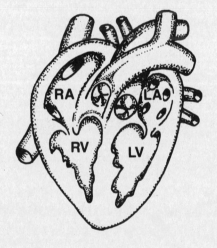
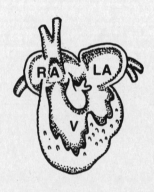

Mammal Heart or Reptile Heart

6. _____   _____

Similarities _____
_____

Differences _____
_____

© Prentice-Hall, Inc.   Human Biology and Health   H ■ 81

   **Platelets
or
Red Blood Cells**

7. _____   _____

Similarities _____
_____

Differences _____
_____

| Anti-A Serum | Anti-B Serum | Anti-A Serum | Anti-B Serum |
| No Clumping | Clumping | Clumping | No Clumping |

**Group AB Blood
or
Group O Blood**

8. _____   _____

Similarities _____
_____

Differences _____
_____

82 ■ H   Human Biology and Health

Name _____ Class _____ Date _____

Activity
Circulatory System

CHAPTER 4

Surveying Risk Factors Related to Heart Disease

Many factors contribute to the occurrence of heart and artery disease. For example, men are more prone to heart attacks than women are. In addition, the following items are often considered to be high risk factors for heart and artery problems: smoking, being over 40 years old, having a stressful job, being overweight, having high blood pressure, not getting enough exercise, and eating foods that are high in fat and cholesterol. Heredity may also play a role.

In this activity you will conduct a survey of adults. You will gather data about the incidence of risk factors related to heart and artery disease.

1. Ask 20 adults the following questions.
   a. Do you smoke?
   b. Does your job cause you a lot of stress?
   c. Are you over 40 years old?
   d. Are you overweight?
   e. Do you exercise on a regular basis?
   Also note whether the person is male or female.

2. Calculate a risk score for each person surveyed. Count 1 point each for a YES answer to questions a, b, c, and d. Count 1 point for a NO answer to question e. Add 1 point if the subject is a male.

3. Complete the following chart to show your results.

Risk Factors for Heart and Artery Disease

Risk Scores	0	1	2	3	4	5	6
Number of People							

4. Use the form on the next page to construct a bar graph of your results.

© Prentice-Hall, Inc.   Human Biology and Health   H ■ 83

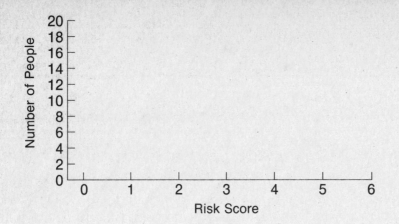

5. What conclusions can you draw from your survey? _____

_____

_____

_____

_____

_____

_____

Name _____ Class _____ Date _____

Laboratory Investigation

CHAPTER 4 ■ Circulatory System

Measuring Your Pulse Rate

Problem
What are the effects of activity on pulse rate?

Materials *(per group)*
clock or watch with a sweep second hand
graph paper

Procedure
1. On a separate sheet of paper, construct a data table similar to the one shown here.
2. To locate your pulse, place the index and middle finger of one hand on your other wrist where it joins the base of your thumb. Move the two fingers slightly until you locate your pulse.
3. To determine your pulse rate, have one member of your group time you for 1 minute. During the 1 minute, count the number of beats in your pulse. Record this number in the data table.
4. Walk in place for 1 minute. Then take your pulse. Record the result.
5. Run in place for 1 minute. Again take your pulse. Record the result.
6. Sit down and rest for 1 minute. Take your pulse. Then take your pulse again after 3 minutes. Record the results in the data table.
7. Use the data to construct a bar graph that compares each activity and the pulse rate you determined.

Observations

	Resting	Walking	Running	Resting After Exercise (1 min)	Resting After Exercise (3 min)
Pulse Rate					

© Prentice-Hall, Inc.

Human Biology and Health   H ■ 85

1. What pulse rate did you record in step 3 of the Procedure? This is called your pulse rate at rest. How does your pulse rate at rest compare with those of the other members of your group? (Do not be alarmed if your pulse rate is somewhat different from those of other students. Individual pulse rates vary.)

2. What effect did walking have on your pulse rate? Running?

3. What effect did resting after running have on your pulse rate?

Analysis and Conclusions

1. What conclusions can you draw from your data?

2. How is pulse rate related to heartbeat?

3. What happens to the blood supply to the muscles during exercise? How is this related to the change in pulse rate?

Answer Key

CHAPTER 4 ■ Circulatory System

Chapter Discovery: Simulating Blood Transfusions

Observations Recipient blood group A can safely receive blood from donor blood groups A and O only. Recipient blood group B can safely receive blood from donor blood groups B and O only. Recipient blood group AB can safely receive blood from all the donor blood groups. Recipient blood group O can only receive blood from donor blood group O.
Analysis and Conclusions 1. A and O
2. B and AB **3.** Group AB. It can mix with all blood groups. **4.** Probably group AB. People with this blood group can only receive blood from others with the same group.
5. Yes.

Discovery Activity: The Heartbeat

Observations and Conclusions 1. A regular "lub-dub" sound at a rate of approximately 72 times per minute **2.** As the heart beats, the atria first contract and then the ventricles, producing the "lub" and "dub" sounds.
3. A stethoscope is used to listen for irregularities in the sound or rate of the heartbeat (or breathing rate) that might indicate health problems.

Problem-Solving Activity: Circulation Calculations

1. Answers will vary depending upon body mass. 45.5 kg of body mass = 3.2 kg of blood mass **2.** 135 L **3.** 300 L; 1500 L; 7200 L
4. 4200 beats; 100,800 beats; 36,792,000 beats **5.** 20,000,000 **6.** Male; 650 times
7. 18,000,000,000 **8.** 4.5 percent
9. Answers will vary (number in school × 0.40). **10.** 80 km/hr

Problem-Solving Activity: Circulation Comparisons

1. Atria, Ventricles: Both are heart chambers. Ventricles are lower heart chambers, are more muscular, and pump blood. Atria are upper heart chambers, have thinner walls, and receive blood. **2.** Vein, Artery: Both carry blood. A vein has thin walls, valves, and generally carries deoxygenated blood. An artery has thicker walls, no valves, and generally carries oxygenated blood. **3.** Red blood cells, White blood cells: Both are human blood cells. Red blood cells are smaller and more numerous, carry oxygen, and have no nuclei. White cells are larger and less numerous, fight diseases, and have nuclei.
4. Group B blood, Group A blood: Both are blood groups in the ABO system and both can be received by Group AB. Group B has B antigens and Group A has A antigens. Each is only compatible with its own group.
5. Plasma, Whole blood: Both are found in blood vessels. Plasma is the liquid part of blood and contains no blood cells. Whole blood contains plasma and blood cells.
6. Mammal heart, Reptile heart: Both are part of the circulatory system of each organism and both pump blood. The mammal heart has 4 chambers. The reptile heart has 3 chambers. **7.** Platelets, Red blood cells: Both make up solid portion of blood. Platelets are cell fragments that aid in blood clotting. Red blood cells are cells that carry oxygen. **8.** Group AB blood, Group O blood: Both are blood groups in the ABO system. Group AB has A and B antigens and is the universal recipient. Group O has no antigens and is the universal donor.

Discovery Activity: Surveying Risk Factors Related to Heart Disease

Answers will vary. Check tables and graphs to see that students have correctly analyzed their data. Students will find that people with the most risk factors will be most prone to heart and artery disease. Impress upon students the importance of good health habits in preventing these diseases.

Laboratory Investigation: Measuring Your Pulse Rate

Observations 1. Pulse rate at rest is typically about 72 beats per minute but may vary widely from student to student. 2. Both running and walking should have increased pulse rate, with running causing a greater increase. 3. Resting should reduce the pulse rate back to or close to its normal rate, depending on the length of the rest period.
Analysis and Conclusions 1. Conclusions may vary but should indicate that activity increases pulse rate, and periods of rest or inactivity tend to minimize pulse rate. 2. The pulse is actually a way to measure the heartbeat because pulse rate and heartbeat are equal. 3. Blood supply to muscles increases during exercise. Because the heart must beat faster to provide an increased blood supply, pulse rate also increases during exercise.

Contents

CHAPTER 5 ■ Respiratory and Excretory Systems

Chapter Discovery
*Breathing and Exercise H91

Chapter Activities
*Discovery Activity: Measuring Pulse and Respiration Rates. H93
*Problem-Solving Activity: What's in the Air That You Breathe?. . . . H95
*Discovery Activity: A Breathing Model. H97
*Discovery Activity: Changes in Pitch. H99
*Discovery Activity: Keeping Cool. H101

Laboratory Investigation Worksheet
Measuring the Volume of Exhaled Air H103
  (**Note:** *This investigation is found on page H126 of the student textbook.*)

Answer Key H105

*Appropriate for cooperative learning

Name _____ Class _____ Date _____

Chapter Discovery — Respiratory and Excretory Systems
CHAPTER 5

Breathing and Exercise

Background Information

Normal breathing rates vary from 12 to 25 times per minute. In this activity you will compare your breathing rate at rest to your breathing rate after exercise.

Problem

How does your breathing rate change when you exercise?

Materials

clock or watch with second hand

Procedure

1. Find your partner's pulse by gently placing two or three fingertips on the inside of your partner's wrist.
2. Sit quietly and breathe normally for 1 minute.
3. Have your partner count the number of breaths you take in 1 minute. This is your normal breathing rate. Record your normal breathing rate in the Data Table.
4. Run in place for 30 seconds. Then sit down and have your partner count the number of breaths you take in 1 minute. Record your breathing rate in the Data Table. Allow your breathing rate to return to normal before continuing on.
5. Run in place for 1 minute. Sit down and have your partner count the number of breaths you take in 1 minute. Record your breathing rate in the Data Table.

Observations
DATA TABLE

Activity	Breathing Rate (breaths per minute)
Resting	
After 30 seconds of exercise	
After 1 minute of exercise	

© Prentice-Hall, Inc.   Human Biology and Health

Analysis and Conclusions

1. How did exercise affect your breathing rate?

   _____
   _____
   _____

2. Can you think of a reason for your answer to question 1?

   _____
   _____
   _____
   _____

3. Did you notice any other way in which your breathing changed with exercise? Give a possible reason for this change.

   _____
   _____
   _____
   _____

4. What other factors besides exercise might influence your normal breathing rate?

   _____
   _____
   _____

Name _____ Class _____ Date _____

Activity
Respiratory and Excretory Systems

CHAPTER 5

Measuring Pulse and Respiration Rates

The pulse can be used as a measure of the heart's activity. The respiration rate indicates the lungs' activity in supplying oxygen to the blood and in removing carbon dioxide from the blood. In this activity you will see if there is a relationship between pulse and respiration rates by comparing these rates when at rest and during physical activity.

Procedure

1. Find your partner's pulse by gently pressing two or three fingertips on the inside of your partner's wrist.
2. Using a clock or watch with a second hand, count the number of the pulse beats in 1 minute. Write this number in the chart.
3. Now count the number of times your partner breathes in 1 minute. Record this information in the chart.
4. To find out your partner's pulse rate and respiration rate after walking, have your partner walk around the room for 1 minute.
5. Then have your partner sit down, and take his or her pulse and respiration rates for 1 minute. Record this information in the chart.
6. After a few minutes' rest, have your partner run in place for 1 minute.
7. Then take his or her pulse and respiration rates for 1 minute. Record this information in the chart.

	Resting	Walking	Running
Pulse rate			
Respiration rate			

Observations and Conclusions

1. What does walking do to the pulse and respiration rates? _____
   _____

2. How does running in place affect pulse and respiration rates? _____
   _____

© Prentice-Hall, Inc.

Human Biology and Health

Name _____ Class _____ Date _____

Activity
Respiratory and Excretory Systems

CHAPTER 5

What's in the Air That You Breathe?

The air that you breathe has various kinds and amounts of pollutants in it.

Procedure

1. To collect evidence of air pollutants, trace the outline of a microscope slide on a sheet of graph paper.
2. Cut out your outline and tape it to one side of the microscope slide. This paper provides a grid for counting dust, pollen, and other particles. Prepare five slides in this manner.
3. Decide where you will place your slides to collect air pollutants. Examples are in a corner of the classroom, in the hall, in the cafeteria, in a tree, outside in an open area.
4. On the back of the graph paper of each slide, write your initials, the date, and where you will put the slide.
5. Now cover the front of each slide with a thin coat of petroleum jelly.
6. Place the slides in the locations you have chosen. Leave them there for at least 24 hours.
7. Collect the slides and count the number of particles in ten of the squares.

Observations and Conclusions

1. How many particles are there on slide 1? _____

   Slide 2? _____

   Slide 3? _____

   Slide 4? _____

   Slide 5? _____

2. a. Which location had the most pollutants? _____

   b. Which location had the fewest pollutants? _____

© Prentice-Hall, Inc.

Human Biology and Health

Activity
Respiratory and Excretory Systems — CHAPTER 5

A Breathing Model

Procedure

1. Insert a glass Y-tube into the opening of a rubber stopper. **CAUTION:** *Be careful when using glass tubing. Follow the instructions of your teacher.*
2. Fasten two balloons of the same size to the two arms of the Y-tube. Use string or rubber bands to keep the balloons in place.
3. Place the rubber stopper in a bell jar.

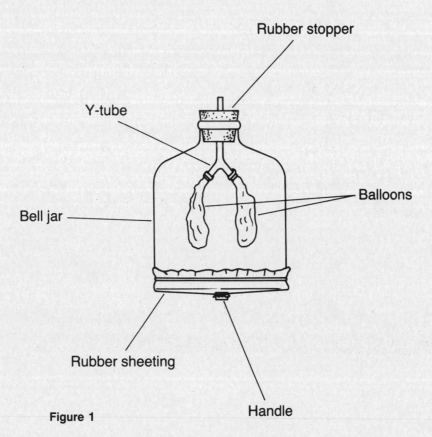

Figure 1

4. Cut a thin circle of rubber sheeting that is large enough to fit over the bottom of the bell jar.
5. In the center of the rubber sheeting, pinch out a piece and insert a small rubber stopper. Using string, tie the rubber stopper off so that you can use it as a place to grasp the rubber sheeting.
6. With the string, tie the rubber sheeting in place over the bottom of the bell jar.
7. Pull down on the rubber sheeting.
8. Push up on the rubber sheeting.

Observations and Conclusions

1. What happens when the rubber sheeting is pulled down? _____

   _____

   Relate this action to the functioning of the respiratory system.

   _____

   _____

   _____

   _____

2. What happens when the rubber sheeting is pushed up? _____

   _____

   Relate this action to the functioning of the respiratory system.

   _____

   _____

   _____

   _____

3. Which respiratory system structure is represented by each of the following?

   a. bell jar _____

   b. balloons _____

   c. top of Y-tube _____

   d. branches of Y-tube _____

   e. rubber sheeting _____

Activity

Respiratory and Excretory Systems

CHAPTER 5

Changes in Pitch

When you speak, the pitch of your voice keeps changing. The changes in pitch add meaning to words. Pitch depends upon the number of times an object such as your vocal cords vibrates. When an object vibrates faster, its pitch becomes higher. When an object vibrates slower, its pitch becomes lower. To change pitch, you must change speed of vibration.

Procedure
1. Obtain a wooden cigar box and remove its cover.
2. Stick two thumbtacks in the box, one on each side.
3. Cut a rubber band open and stretch it across the open side of the box.
4. Attach each end to a thumbtack.
5. Pluck the rubber band and listen to the sound.
6. Now pull the rubber band tighter by moving the thumbtacks farther back on the box. Again pluck the rubber band and listen to the sound.

Did the rubber band vibrate faster or slower when it was pulled tighter? _____

Was the pitch higher or lower when the rubber band was pulled tighter? _____

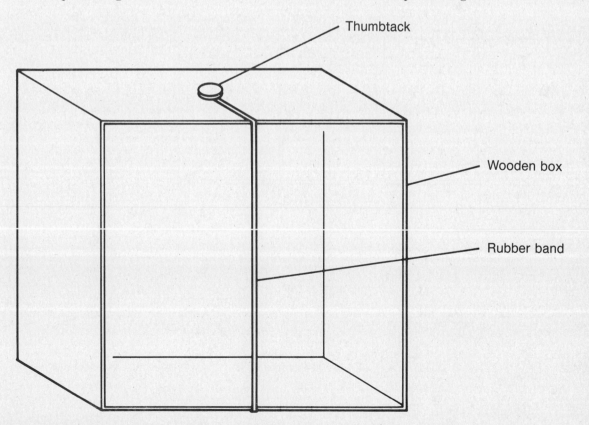

© Prentice-Hall, Inc.

Human Biology and Health

7. Remove the rubber band and thumbtacks from the wooden box.
8. Now stick three thumbtacks on each side of the box, all the same distance from the open side.
9. Cut three rubber bands of different thicknesses open and stretch them across the open side of the box by attaching them to the thumbtacks.
10. Pluck each band and listen to the sound.

Which rubber band vibrates fastest? _____

Which rubber band has the highest pitch? _____

Explain the relationship between pitch and vibration. _____

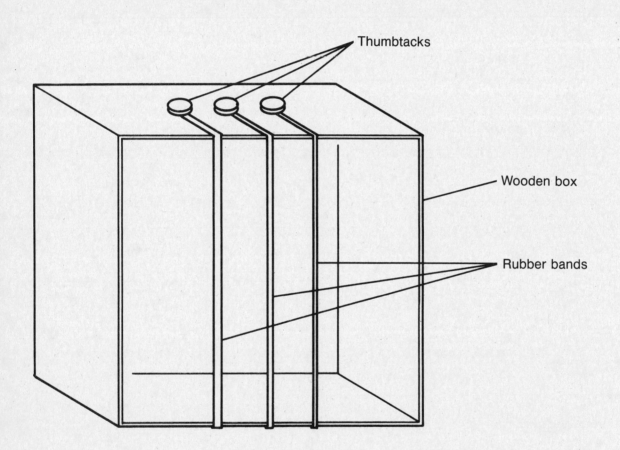

Name _____ Class _____ Date _____

Activity

Respiratory and Excretory Systems

CHAPTER 5

Keeping Cool

Procedure

1. Wet a cotton swab with rubbing alcohol and rub it on the back of your hand. **CAUTION:** *Rubbing alcohol is poisonous if ingested.*
2. Repeat step 1, using water instead of rubbing alcohol. Be sure the water is at room temperature.
3. Observe what happens.
4. Obtain three thermometers. Label them 1 to 3. Wrap one piece of cotton around thermometers 1 and 2. Record the temperature of each thermometer in the chart below.
5. Soak the cotton on thermometer 1 with water. Record the temperature after 1 minute.
6. Soak the cotton on thermometer 2 with rubbing alcohol. Record the temperature after 1 minute.
7. Keep thermometer 3 dry. Record its temperature.

Thermometer	Temperature (°C) at Start of Activity	Temperature (°C) After One Minute
1		
2		
3		

Observations and Conclusions

1. What happened when the water was placed on your skin? _____

_____

_____

2. What happened when the rubbing alcohol was placed on your skin? _____

_____

© Prentice-Hall, Inc.

Human Biology and Health

3. What caused the sensation in questions 1 and 2? _____

4. Which of the three thermometers showed the lowest reading? _____
   Explain. _____

5. What was the purpose of thermometer 3? _____

6. What is the relationship between the skin and the thermometers? _____

Laboratory Investigation

CHAPTER 5 ■ Respiratory and Excretory Systems

Measuring the Volume of Exhaled Air

Problem
What is the volume of exhaled air?

Materials *(per group)*
glass-marking pencil
spirometer
red vegetable coloring
paper towel
graduated cylinder

Procedure

1. Obtain a spirometer. A spirometer is an instrument that is used to measure the volume of air that the lungs can hold.

2. Fill the plastic bottle four-fifths full of water. Add several drops of vegetable coloring to the water. With the glass-marking pencil, mark the level of the water.

3. Reattach the rubber tubing as shown in the diagram.

4. Cover the lower part of the shorter length of rubber tubing with the paper towel by wrapping the towel around it. This is the part of the rubber tubing that you will need to place your mouth against. **Note:** *Your mouth should not come in contact with the rubber tubing itself, only with the paper towel.*

5. After inhaling normally, exhale normally into the shorter length of rubber tubing.

6. The exhaled air will cause an equal volume of water to move through the other length of tubing into the graduated cylinder. Record the volume of this water in milliliters in a data table.

7. Pour the colored water from the cylinder back into the plastic bottle.

8. Repeat steps 3 through 7 two more times. Record the results in your data table. Calculate the average of the three readings.

9. Run in place for 2 minutes and exhale into the rubber tubing. Record the volume of the water in the graduated cylinder.

10. Rest for a few minutes until your breathing returns to normal. Then repeat step 9 two more times and record the results. Calculate the average of the three readings.

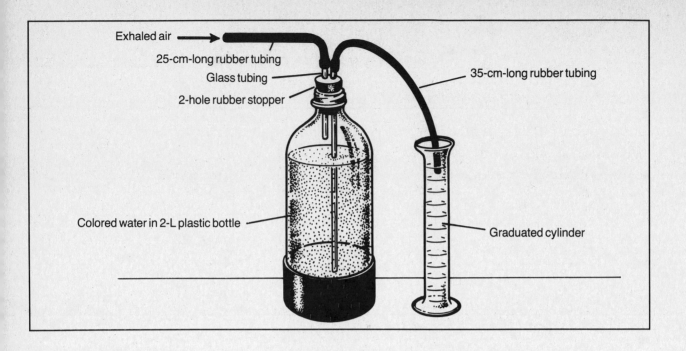

Observations
How does your average volume of exhaled air before exercise compare to your average volume of exhaled air after exercise?

_____

_____

Analysis and Conclusions
1. Why is it important to measure the volume of exhaled air three times?

   _____

   _____

2. Explain how exercise affects the volume of exhaled air.

   _____

   _____

3. **On Your Own** Describe how you could determine the volume of air you exhale in a minute.

   _____

   _____

   _____

   _____

Answer Key

CHAPTER 5 ■ Respiratory and Excretory Systems

Chapter Discovery: Breathing and Exercise

Observations Check table to see if students have logical data. Students will find that breathing rates will increase as more exercise is performed. **Analysis and Conclusions 1.** Breathing rate should increase with exercise; it probably will be greater after 1 minute of exercise than after 30 seconds. **2.** Possible answer: The body needs more energy during exercise, and energy is obtained by the process of respiration, which requires oxygen. An increased breathing rate allows more oxygen to enter the body. **3.** Students will probably notice that their breathing becomes deeper and harder after exercise. A possible reason for this is that harder breathing allows more oxygen to enter the lungs with each breath. **4.** Possible answers may include: being nervous or frightened might increase your breathing rate; being very sleepy or relaxed might slow it down; very cold or very hot weather might affect breathing rate; certain medicines or illnesses could also influence your breathing rate.

Discovery Activity: Measuring Pulse and Respiration Rates

Observations and Conclusions Be sure students do not use their thumbs to take the pulse of other students. The pronounced pulse in the thumb may interfere with detection of the pulse in the wrist. Answers in the chart will vary. **1.** Walking causes a moderate increase in both pulse and respiration rates. **2.** Running causes an even greater increase in the rates.

Problem-Solving Activity: What's in the Air That You Breathe?

Observations and Conclusions Answers will vary. But students should discover that industrial areas will have more pollution than those areas that have very little or no industry.

Discovery Activity: A Breathing Model

Observations and Conclusions 1. When the rubber sheeting is pulled down, the volume inside the bell jar (chest cavity) increases, decreasing the air pressure inside. As a result, the balloons (lungs) inflate. **2.** When the rubber sheeting is pushed upward, the volume inside the bell jar (chest cavity) decreases, increasing the air pressure inside. As a result, the balloons (lungs) deflate. **3.** a. chest cavity b. lungs c. trachea d. bronchi e. diaphragm

Discovery Activity: Changes in Pitch

The rubber band vibrates faster when it is pulled tighter. The pitch becomes higher when the rubber band is pulled tighter. Of the three rubber bands, the thinnest vibrates fastest and has the highest pitch. The faster the vibration, the higher the pitch.

Discovery Activity: Keeping Cool

Observations and Conclusions 1. The area of the skin that came in contact with the water felt cool and wet. **2.** The area of the skin that came in contact with the rubbing alcohol felt much cooler than the water. **3.** The temperature receptors in the skin were stimulated by the rapid evaporation of the rubbing alcohol, causing a cooling effect. **4.** Thermometer 2 had the lowest reading because the rapid evaporation of the alcohol causes a cooling effect. As a result, the temperature goes down. **5.** Thermometer 3 acts as a control. **6.** The skin feels the cooling effect of the rubbing alcohol, while the thermometer shows the actual drop in temperature.

Laboratory Investigation: Measuring the Volume of Exhaled Air

Observations For most students, exercise will increase the need for oxygen, and the body will compensate by inhaling and exhaling

faster and deeper. Thus, the average amount of exhaled air should be greater after exercise.   **Analysis and Conclusions**
1. Three measurements will ensure more accurate data, as each exhalation may be slightly different from the next. Three trials will give a more accurate average.
2. Exercise increases the volume of exhaled air. This is due in part because exercise causes an increase in the amount of carbon dioxide wastes produced by the body, which, in turn, causes the body to exhale more deeply to eliminate this excess carbon dioxide.
3. Count the number of breaths normally taken per minute. Multiply the number of breaths per minute by the average volume of exhaled air before exercise. This would give the volume of air exhaled in a minute.

Contents

CHAPTER 6 ■ Nervous and Endocrine Systems

Chapter Discovery
*Creating a Feedback Mechanism H109

Chapter Activities
*Discovery Activity: The Blind Spot H113
*Activity: H.E.L.P. From Your Endocrine System H115
*Discovery Activity: The Senses of Taste and Smell H117
*Problem-Solving Activity: Endocrine Riddles H119
*Activity: The Magic Square H121
*Discovery Activity: Your Senses H123
*Discovery Activity: Determining Reaction Time H127

Laboratory Investigation Worksheet
Locating Touch Receptors H129
   (**Note:** *This investigation is found on page H162 of the student textbook.*)

Answer Key H131

*Appropriate for cooperative learning

Name _____ Class _____ Date _____

CHAPTER 6

Chapter Discovery Nervous and Endocrine Systems

Creating a Feedback Mechanism

Background Information

Hormones are chemical messengers that help regulate certain body activities. They are produced in your body by endocrine glands. The amount of each type of hormone in your body is controlled by a process called a feedback mechanism. In this activity you will discover how a feedback mechanism works.

Materials
2 glass jars or beakers
glass-marking pencil
metric ruler
red and blue food coloring
2 medicine droppers
2 paper cups

Procedure
1. Label the glass jars A and B.
2. Place the metric ruler next to jar B as shown in Figure 1. Then with the glass-marking pencil place a mark on the jar 1 cm from the bottom. Make five more marks above the first one at 1-cm intervals. Label the marks 1 through 6 to indicate the number of centimeters from the bottom of the jar.

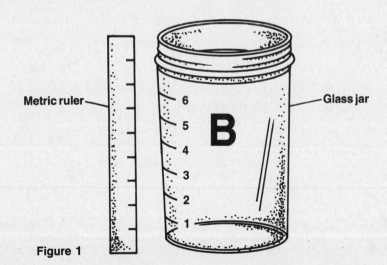

Figure 1

3. Label one paper cup Hormone A and the other paper cup Hormone B.
4. Fill both paper cups about three-fourths full with water. Add several drops of red food coloring to the cup labeled Hormone A and several drops of blue food coloring to the cup labeled Hormone B.

© Prentice-Hall, Inc. Human Biology and Health H ■ 109

5. Place the glass jars a short distance apart on a flat surface such as a desk or table. Place the cup labeled Hormone A next to jar A and the cup labeled Hormone B next to jar B. See Figure 2.

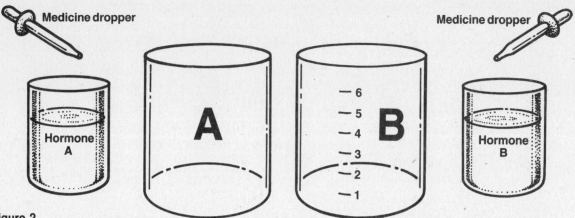

Figure 2

6. Use one of the medicine droppers to add a dropperful of red-colored water in the cup labeled Hormone A to jar A. Then use the other medicine dropper to add a dropperful of blue-colored water in the cup labeled Hormone B to jar B.

7. Add another dropperful of red-colored water to jar A. Add another dropperful of blue-colored water to jar B.

8. Observe the level of the water in jar B. If it is less than 4 cm, repeat step 7. If it is 4 cm or more, go to step 9.

9. Remove a dropperful of blue-colored water from jar B and return it to the paper cup.

10. Observe the level of the water in jar B. If it is more than 2 cm, repeat step 9. If it is 2 cm or less, repeat steps 6 through 8. When you reach step 9 stop.

Critical Thinking and Application

Imagine that jars A and B represent the levels of Hormones A and B in your body. Also imagine that each dropperful of red-colored water added to jar A represents the production of Hormone A, and each dropperful of blue-colored water added to jar B represents the production of Hormone B.

1. What happens when there is an increase in the production of Hormone A?

_____

_____

2. What caused the production of Hormone A to continue to increase?

_____

_____

Name _____ Class _____ Date _____

3. What caused the production of Hormone A to stop increasing?

_____

_____

4. What caused the production of Hormone A to start increasing again?

_____

_____

5. What is the purpose of Hormone A?

_____

_____

6. What regulates the production of Hormone A?

_____

_____

7. What is the relationship between the production of Hormone A and Hormone B in the body?

_____

_____

_____

_____

8. Use a flow chart to show the relationship that you described in question 7.

© Prentice-Hall, Inc. Human Biology and Health H ■ 111

Name _____ Class _____ Date _____

Activity

Nervous and Endocrine Systems

CHAPTER 6

The Blind Spot

Because no nerve receptors are found in the area where the optic nerve is attached to the retina, this area is called the blind spot. To locate your blind spot, look at the drawings below. Center and hold this page about 50 cm in front of your eyes so that the + is on your left. Close your right eye and focus your left eye on the ●. Slowly move the page toward your eyes while looking at the ●.

1. Does the + disappear? _____

   This is known as the blind spot.

2. Using a meterstick, have your partner measure the distance between your eyes and the page. At what distance from your eyes do you reach your blind spot?

   _____

3. Continue to move the page toward you. What happens to the +? _____

4. Repeat the process using your right eye and focusing on the +. Does the ● disappear? _____

5. At what distance from your eyes do you reach your blind spot? _____

6. What happens to the ● as you continue to move the page toward you? _____

   _____

© Prentice-Hall, Inc.

Human Biology and Health   H ■ 113

Name _____ Class _____ Date _____

Activity
Nervous and Endocrine Systems

CHAPTER 6

H.E.L.P. From Your Endocrine System

Complete the following chart by filling in the blank boxes with the information that is needed. If you need help, consult your textbook or other reference books.

Hormone(s)	Endocrine Gland	Location of Gland	Purpose of Hormone
1.	2.	Neck, below voice box	3.
4.	Pituitary	5.	6.
Parathyroid hormone	7.	8.	9.
10.	Adrenals	11.	12.
13.	14.	15.	Regulates sugar level in blood

© Prentice-Hall, Inc.

Human Biology and Health H ■ 115

Name _____ Class _____ Date _____

Activity

Nervous and Endocrine Systems

CHAPTER 6

The Senses of Taste and Smell

1. Using a knife, cut a raw potato, apple, pear, and onion into small cubes. **CAUTION:** *Be very careful when using a knife.*
2. Working in pairs, blindfold one person.
3. While the blindfolded person pinches his or her nostrils closed, the other person should use a spoon to place a small cube of one of the foods into the blindfolded person's mouth. **CAUTION:** *Do not put anything other than the foods to be tested into each other's mouths.*
4. The blindfolded person should chew the food and then carefully spit it out into a paper towel. Dispose of the paper towel.
5. Identify the food and record the information in the chart on the left.
6. Repeat steps 3 to 5 for the other two foods.
7. Repeat the activity while keeping the blindfold on, but do not pinch the nostrils shut. Record your information in the chart on the right.

Nostrils Closed

Food	Identification of Food
Potato	
Apple	
Pear	
Onion	

Nostrils Open

Food	Identification of Food
Potato	
Apple	
Pear	
Onion	

Can the foods be identified by taste alone? Explain. _____
_____
_____
_____

© Prentice-Hall, Inc.

Human Biology and Health

Activity

Nervous and Endocrine Systems

CHAPTER 6

Endocrine Riddles

Read each of the following descriptions. Then, fill in the name of the correct endocrine gland, hormone, or hormonal disorder that is described.

_____ 1. In 1939, at the age of 20, I stood 270 cm tall and wore size 37 shoes! The hormonal disorder from which I suffer is known as giantism. Too much of what hormone produced my condition?

_____ 2. My thyroid gland did not function properly when I was a baby. As a result, both my physical and mental growth were not normal. Although I am 18 years old, I am only 130 cm tall. I am also mentally retarded. What hormone did I lack when I was growing?

_____ 3. I hate seafood! I also use regular salt rather than iodized salt. As a result of too little iodine in my diet, I developed a large lump on my neck. This condition is called goiter. An enlargement of what gland in my neck caused the goiter?

_____ 4. Sometimes I feel like a pincushion because I have had so many needles stuck in me! I have diabetes and need an injection of a certain hormone every day to regulate the sugar level in my blood. What is the name of the hormone that my body does not adequately produce?

_____ 5. I work in the circus as a clown. I am 35 years old and am only 114 cm tall. Although I am small, I have normal intelligence. As a child, one of my glands did not function properly. Too little human growth hormone was secreted, which caused my stunted growth. Which gland caused this problem?

_____ 6. I just can't seem to lose weight! The doctor tells me that my metabolic rate is very low. I feel very tired and sluggish all the time. My body may be producing too little of what hormone?

© Prentice-Hall, Inc.

Human Biology and Health   H ■ 119

_____ 7. I recently had my picture in the newspaper. The reporter called me Superman. I was involved in an automobile accident in which my friend was trapped underneath the car. Somehow I was able to lift the car in order to rescue him. I don't know how I did it! What hormone must have helped me have this extraordinary strength?

_____ 8. Ten years ago, at age 30, I was 176 cm and had a mass of 63.5 kg. Then my pituitary gland began secreting too much of a certain hormone. I did not get taller because my bones were already fully formed. Instead, my nose became much larger and my other facial bones also thickened. I now have a mass of 90.72 kg and look very different! What hormone was overproduced to cause this condition?

_____ 9. Because I have diabetes, I must watch what I eat very carefully. The doctor told me not to eat too many sweet things; and, oh, how I miss candy bars! I seem to be thirsty a lot, too. What gland is not producing the right amount of insulin?

Activity

Nervous and Endocrine Systems

CHAPTER 6

The Magic Square

Match the definitions in column 1 with the correct word or words in column 2. Write the number of the word in the magic square. The magic number can then be found by adding the columns either vertically or horizontally. All rows and columns in the magic square will add up to the same number. **Note:** *One item in column 2 will not be used.*

Column 1
A. Body's network of chemical control
B. Chemicals secreted by endocrine system
C. Hormone that produces fight-or-flight reaction
D. Pumps chemicals through ducts into nearby organs
E. Controls body temperature, appetite, and sleep
F. Produces HGH, human growth hormone
G. Lack of this hormone may cause dwarfism
H. Too much HGH may produce this condition
I. H-shaped gland that produces calcitonin and thyroxine
J. Hormone that helps control metabolism
K. Four small glands found on or in the thyroid
L. Hormone that controls level of calcium in blood and helps maintain balance of phosphorus
M. Glands located on top of kidneys
N. Gland that functions as both endocrine and exocrine gland and is located below stomach
O. Hormone that reduces level of sugar in blood
P. Hormone that causes a rise in level of sugar in blood

Column 2
1. Endocrine system
2. Parathyroid hormone
3. Insulin
4. Pituitary
5. Parathyroids
6. Hormones
7. Hypothalamus
8. Glucagon
9. Pancreas
10. HGH
11. Exocrine glands
12. Thyroid
13. Calcitonin
14. Giantism
15. Adrenal glands
16. Thyroxine
17. Adrenaline

Magic Square

A	B	C	D
E	F	G	H
I	J	K	L
M	N	O	P

Magic Number: _____

Name _____ Class _____ Date _____

Activity

Nervous and Endocrine Systems

CHAPTER 6

Your Senses

In this activity you will test your tongue's sensitivity to different tastes, map your skin's response to cold and heat, and test your ability to determine changes in mass.

A. Taste Receptors on the Tongue

Humans can distinguish four kinds of tastes: sweet, salty, sour, and bitter. Different parts of the tongue are more sensitive to one taste than to the other tastes. Working with a partner, obtain four solutions: sugar (sweet); salt (salty); lemon (sour); and quinine (bitter).

1. Place a clean cotton swab into each of the solutions. You will also need a paper cup full of drinking water.
2. Have your partner rinse out his or her mouth with the drinking water.
3. Touch the end of the swab to one of the five regions of his or her tongue shown in the drawing below.

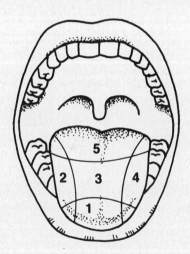

4. Have your partner raise his or her hand if the taste of the substance is detected on any of the areas touched with the moistened cotton swab.
5. Also have your partner indicate the specific taste.
6. Record the results in the chart on the next page. Use a plus (+) sign for a positive detection and a minus (−) sign for a negative detection.
7. Repeat this procedure using each of the remaining solutions.

© Prentice-Hall, Inc.

Human Biology and Health   H ■ 123

Tongue Region	Salty	Sweet	Sour	Bitter
Front (1)				
Side (2)				
Middle (3)				
Side (4)				
Back (5)				

B. Temperature Receptors on the Skin

1. On a separate sheet of paper, trace your partner's hand, palm side down.
2. Blindfold your partner.
3. Chill a glass stirring rod in ice water and touch it to the skin of the back of your partner's hand.
4. Mark a red "x" on the tracing of your partner's hand where your partner sensed a feeling of coldness. **Note:** *Be sure your partner reports sensations of cold, not touch.* Test at least 20 points on your partner's hand.
5. Repeat the procedure using a glass rod warmed in hot water. **CAUTION:** *The water should not be too hot.*
6. Mark a blue "o" on the tracing of your partner's hand to show each point where your partner felt a sensation of heat. Test at least 20 points on your partner's hand.

Where on the hand are the receptors for temperature located? _____

_____

Where does your partner have sensations of cold? _____

Of heat? _____

_____

C. Determining Changes in Mass

1. Place 50 mL of water into each of two 100-mL beakers.
2. Label one beaker A and the other B.
3. Using a balance, find the mass of each beaker with water.

What is the mass of beaker A? _____

Of beaker B? _____

Name _____ Class _____ Date _____

4. Have your partner close his or her eyes and place a beaker in each of his or her hands.

5. Then add small amounts of water to beaker A until your partner senses a difference in mass.

   How many milliliters of water did you add to beaker A? _____

6. To determine the difference in the mass of beaker A after adding the water, place it on a balance. From this mass, subtract the mass of beaker A containing 50 mL of water.

   What is the difference in this mass? _____

Name _____ Class _____ Date _____

Activity

Nervous and Endocrine Systems

CHAPTER 6

Determining Reaction Time

Purpose
  In this activity you will determine your relative length of reaction time—the length of time that elapses between your perceiving a change in your environment and your reacting to that change—and compare it with that of your classmates.

Materials *(per pair of students)*
Metric ruler

Procedure
1. Working with a partner, have him or her hold a ruler vertically above a table.
2. Position your thumb and forefinger around the zero end of the ruler, but not touching it, as shown in the diagram below.

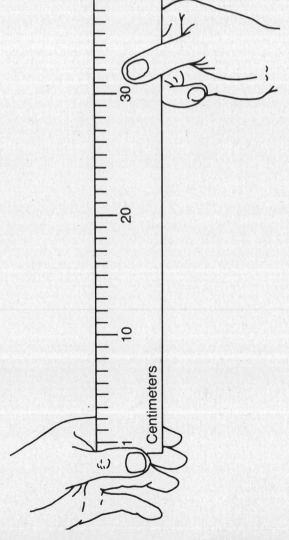

© Prentice-Hall, Inc.

Human Biology and Health   H ■ 127

3. Your partner may drop the ruler whenever he or she chooses. Moving only your thumb and forefinger, not your hand, you must catch the ruler as soon as it falls. In the Data Table, record the distance, in centimeters, from zero that the ruler falls until it is caught.
4. Repeat steps 1 and 3 four more times.
5. To obtain an average distance, add the five distances together and divide this sum by five. Record the average distance in the Data Table.

Observations and Conclusions
Data Table

Trial	Distance
1	
2	
3	
4	
5	
Average	

1. Why does measuring the distance the ruler falls give a relative measure of reaction time?

2. Compare the reaction times of all the students. What can you conclude?

Name _____ Class _____ Date _____

Laboratory Investigation

CHAPTER 6 ■ Nervous and Endocrine Systems

Locating Touch Receptors

Problem
Where are the touch receptors located on the body?

Materials *(per pair of students)*
scissors
metric ruler
blindfold
9 straight pins
piece of cardboard (6 cm × 10 cm)

Procedure

1. Using the scissors, cut the piece of cardboard into five rectangles each measuring 6 cm × 2 cm.
2. Into one cardboard rectangle, insert two straight pins 5 mm apart. Into the second cardboard rectangle, insert two pins 1 cm apart. Insert two pins 2 cm apart into the third rectangle. Insert two pins 3 cm apart into the fourth rectangle. In the center of the remaining cardboard rectangle, insert one pin.

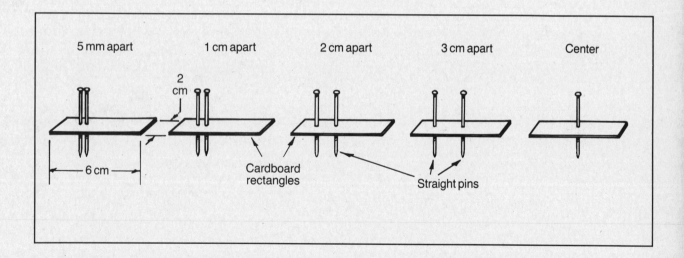

3. Construct a data table in which the pin positions on the cardboard appear across the top of the table.
4. Blindfold your partner.

© Prentice-Hall, Inc.

Human Biology and Health   H ■ 129

5. Using the cardboard rectangle with the straight pins 5 mm apart, carefully touch the palm surface of your partner's fingertip, palm of the hand, back of the hand, back of the neck, and inside of the forearm. **CAUTION:** *Do not apply pressure when touching your partner's skin.* In the data table, list each of these body parts.
6. If your partner feels two points in any of the areas that you touch, place the number 2 in the appropriate place in the data table. If your partner feels only one point, place the number 1 in the data table.
7. Repeat steps 5 and 6 with the remaining cardboard rectangles.
8. Reverse roles with your partner and repeat the investigation.

Observations
On which part of the body did you feel the most sensation? The least?

_____

_____

Analysis and Conclusions
1. Which part of the body that you tested had the most touch receptors? The fewest? How do you know?

_____

_____

2. Rank the body parts in order from the most to the least sensitive.

_____

_____

3. What do the answers to questions 1 and 2 indicate about the distribution of touch receptors in the skin?

_____

_____

4. **On Your Own** Obtain a variety of objects. Blindfold your partner and hand one of the objects to your partner. Have your partner describe how the object feels. Your partner is not to name the object. Record the description along with the name of the object. Repeat the investigation for each object. Reverse roles and repeat the investigation. How well were you and your partner able to "observe" with the senses of touch?

_____

_____

Answer Key

CHAPTER 6 ■ Nervous and Endocrine Systems

Chapter Discovery: Creating a Feedback Mechanism
Critical Thinking and Application
1. There is an increase in the production of Hormone B. 2. The level of Hormone B dropped below 4 cm. 3. The level of Hormone B rose to 4 cm. 4. The level of Hormone B dropped to 2 cm or lower. 5. It stimulates the production of Hormone A. 6. The level of Hormone B 7. When the level of Hormone B rises above a certain level, the production of Hormone A ceases. When the level of Hormone B falls below a certain level, the production of Hormone A starts. 8. Check flow charts to see if they reflect students' descriptions in question 7.

Discovery Activity: The Blind Spot
1. Yes. 2. Answers will vary. 3. It reappears. 4. Yes. 5. Answers will vary. 6. It reappears.

Activity: H.E.L.P. From Your Endocrine System
1. calcitonin; thyroxine 2. thyroid 3. controls level of calcium and phosphorus; regulates metabolism 4. human growth hormone; gonadotropic and lactogenic hormones; ACTH; oxytocin; vasopressin; thyrotropic hormone 5. base of brain below hypothalamus 6. regulates growth; controls blood pressure and water balance; stimulates other glands 7. parathyroid 8. on or in thyroid 9. controls level of calcium in blood; helps maintain balance of phosphorus 10. adrenaline; cortin 11. above kidneys 12. speeds up heart and other body processes; regulates salt and water balance 13. insulin; glucagon 14. pancreas 15. below stomach

Discovery Activity: The Senses of Taste and Smell
Students will discover that they will probably not be able to identify the foods when their nostrils are pinched closed. The sense of smell is needed in order to taste food.

Problem-Solving Activity: Endocrine Riddles
1. human growth hormone 2. thyroxine 3. thyroid 4. insulin 5. pituitary 6. thyroxine 7. adrenaline 8. human growth hormone 9. pancreas

Activity: The Magic Square
A. 1 B. 6 C. 17 D. 11 E. 7 F. 4 G. 10 H. 14 I. 12 J. 16 K. 5 L. 2 M. 15 N. 9 O. 3 P. 8 The magic number is 35.

Discovery Activity: Your Senses
A. **Front (1)** salty; sweet **Side (2)** salty; sour **Middle (3)** no taste **Side (4)** salty; sour **Back (5)** bitter **B, C.** Answers will vary.

Discovery Activity: Determining Reaction Time
Observations and Conclusions Check Data Table. 1. The farther the ruler falls, the longer the reaction time. 2. There will be some distribution of values because reaction times vary.

Laboratory Investigation: Locating Touch Receptors
Observations The most sensation is felt on the fingers, especially the fingertips. The least

sensation is felt on the forearm. **Analysis and Conclusions 1.** The fingers had the most touch receptors because it was easiest to feel the touch of two pins 5 mm apart. For the body part with the fewest touch receptors, answers will vary. The forearm may have the smallest number of receptors because it is more difficult to distinguish between the pins that are 5 mm apart and the one single pin. **2.** Answers will vary, but the fingers and hand should be first, and the forearm will likely be last. **3.** They are not distributed evenly. The high concentration of touch receptors in the hand and fingers makes the hand more efficient for grasping, holding, and touching objects. **4.** Answers will vary. After students have completed the activity, you may want to create a class list on the chalkboard of all the adjectives students used to describe the "feel" of objects.

Contents

CHAPTER 7 ■ Reproduction and Development

Chapter Discovery
*Baby in the Making H135

Chapter Activities
*Discovery Activity: Name It! H139
*Activity: Stages of Human Development H141
*Problem-Solving Activity: Male or Female? H143

Laboratory Investigation Worksheet
How Many Offspring? H145
(**Note:** *This investigation is found on page H186 of the student textbook.*)

Answer Key H147

*Appropriate for cooperative learning

Name _____ Class _____ Date _____

Chapter Discovery — Reproduction and Development

CHAPTER 7

Baby in the Making

Listed below are some important events in the life of a developing human baby. Following the list is a time line that shows the nine months of human pregnancy. Study the list carefully, then place each event on the numbered line on the time line. For now, do not use reference sources. Keep your time line until you have finished studying this chapter. Then compare your answers to what you have learned about the development of a human baby. If at this time you still have questions, then consult reference sources.

Events in the Life of a Developing Baby
- nerve chord forms
- upper and lower eyelids separate and eyelashes form
- three layers of cells form: endoderm, mesoderm, and ectoderm
- facial muscles move
- tubelike structure that will become the heart begins to beat
- arms and legs move
- heartbeat can be heard with the aid of a stethoscope
- developing baby can survive on its own if necessary
- four chambers of the heart form
- sucking reflex begins
- finger rays appear
- all individual organs and body systems have appeared
- nervous system begins to form
- tube that will give rise to digestive system forms
- fingers and toes form
- arm and leg buds appear
- eyes open
- lung development becomes complete
- eye patches appear
- fingers can make grasping movements

© Prentice-Hall, Inc.

Human Biology and Health H ■ 135

TIME LINE

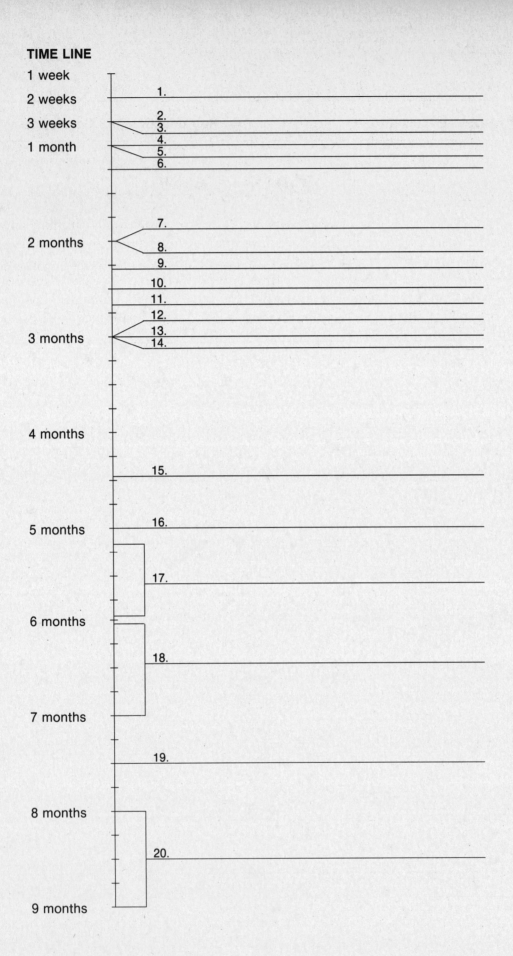

136 ■ H   Human Biology and Health

Name _____ Class _____ Date _____

Critical Thinking and Application

1. What factors did you consider as you decided where to place each item on the time line?

   _____
   _____
   _____
   _____
   _____

2. Which events were the most difficult for you to place?

   _____
   _____
   _____
   _____

3. What questions came to mind as you considered the development of a human baby?

   _____
   _____
   _____
   _____
   _____

Name _____ Class _____ Date _____

Activity

Reproduction and Development

CHAPTER 7

Name It!

You probably are well aware that a male chicken is called a rooster, a female is called a hen, and the young chicken is called a chick. Can you name the male, female, and young of the following animals? If you have difficulty, consult a reference book.

Animal	Male	Female	Young
1. Deer			
2. Duck			
3. Elephant			
4. Fox			
5. Goose			
6. Sheep			
7. Swine			
8. Whale			
9. Swan			
10. Rabbit			
11. Kangaroo			
12. Goat			
13. Lion			
14. Seal			
15. Cattle			

© Prentice-Hall, Inc.

Human Biology and Health

Activity

Reproduction and Development

CHAPTER 7

Stages of Human Development

1. The pictures below show various stages in the development of a human baby. You know that life begins with a fertilized egg that divides many, many times. An embryo is formed from these cells. The embryo develops into a fetus. Eventually, a baby is born. Can you put these pictures in order from fertilization to birth? Put a 1 below the picture that comes first, a 2 below the picture that comes next, and so on.

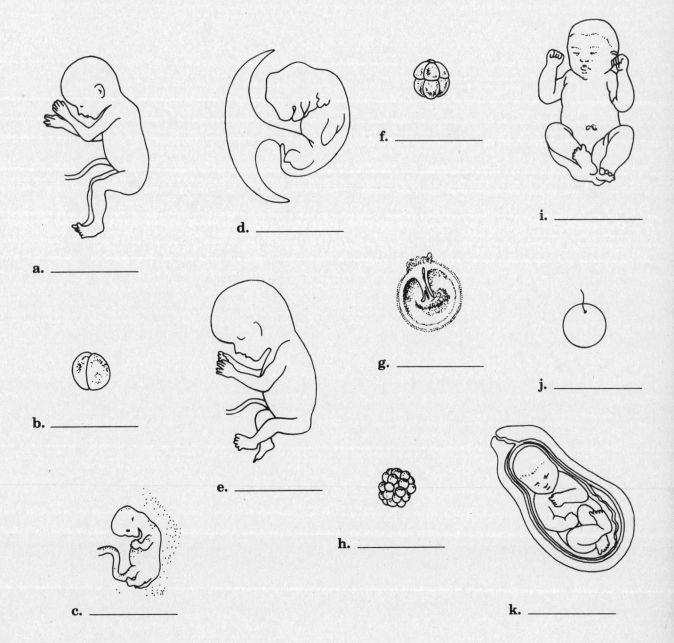

2. About how long does this process of development take? _____

3. Find the umbilical cord in one of the drawings and label it.

4. What structure is connected to the embryo by the umbilical cord? _____

5. What are the strong muscular contractions that push the baby out of the uterus called? _____

6. Find the uterus in one of the drawings and label it.

Name _____ Class _____ Date _____

Activity
Reproduction and Development

CHAPTER 7

Male or Female?

The type of chromosomes carried in the human sperm cell determine the sex of the baby that will develop from the fertilized egg. Each egg carries an X chromosome. A sperm may carry either an X or a Y chromosome. If an X-bearing sperm fertilizes the egg, a female results (XX). A male is formed when a Y-bearing sperm fertilizes the egg (XY).

When biologists wish to study human chromosomes, they sometimes photograph the chromosomes through a microscope. After these photographs are enlarged, the chromosomes are cut out and arranged in pairs. The resulting product is called a karyotype.

Human Male Karyotype

Human Female Karyotype

1. Construct a karyotype from the chromosomes shown on the next page. Cut out the individual (doubled) chromosomes and match up the pairs. Then arrange the pairs as shown above. The last step is to number the pairs.

2. How many pairs of chromosomes are there in human cells? _____

3. Is the karyotype you made that of a male or a female? _____

   How can you tell? _____

   _____

© Prentice-Hall, Inc.   Human Biology and Health   H ■ 143

144 ■ H Human Biology and Health

Name _____ Class _____ Date _____

Laboratory Investigation

CHAPTER 7 ■ Reproduction and Development

How Many Offspring?

Problem
How do the length of gestation, number of offspring per birth, age of puberty, and life span of various mammals compare?

Materials *(per group)*
graph paper
colored pencils

Procedure
1. Study the chart, which shows the length of gestation (pregnancy), the average number of offspring per birth, the average age of puberty, and the average life span of certain mammals.
2. Construct a bar graph that shows the length of gestation for each mammal.
3. Construct another bar graph that shows the average number of offspring of each mammal.
4. Construct a third bar graph that shows the life span of each mammal. Color the portion of the bar that shows the length of childhood, or time from birth to puberty.

Mammal	Gestation Period (days)	Number of Offspring per Birth	Age at Puberty	Life Span (years)
Opossum	12	13	8 months	2
House mouse	20	6	2 months	3
Rabbit	30	4	4 months	5
Dog	61	7	7 months	15
Lion	108	3	2 years	23
Rhesus monkey	175	1	3 years	20
Human	280	1	13 years	74
Horse	330	1	1.5 years	25

© Prentice-Hall, Inc.

Human Biology and Health

Observations

1. Which of the mammals has the longest gestation period?

2. Which mammal has the largest number of offspring per birth?

3. Which of the mammals has the shortest life span?

4. Which mammal takes longer to reach puberty than Rhesus monkeys?

Analysis and Conclusions

1. What general conclusions can you draw after studying the graphs you have made?

2. If a mouse produces five litters per year, how many mice does the average female mouse produce in a lifetime?

3. Of all the mammals listed, which care for their young for the longest period of time after birth? Why do you think this is the case?

4. **On Your Own** Gather the same kinds of data for five additional mammals. Add these data to your existing graphs. How do the five new mammals compare with those provided in this investigation?

Answer Key

CHAPTER 7 ■ Reproduction and Development

Chapter Discovery: Baby in the Making
TIME LINE 1. Three layers of cells form
2. Nervous system begins to form **3.** Tube that will become digestive system forms
4. Tubelike structure that will become heart begins to beat **5.** Arm and leg buds appear
6. Finger rays appear **7.** Nerve chord forms **8.** Four chambers of heart form
9. Facial muscles move **10.** Arms and legs move **11.** Sucking reflex begins **12.** Eye patches appear **13.** Fingers and toes form
14. All individual organs and body systems have appeared **15.** Fingers can make grasping movements **16.** Heartbeat can be heard with the aid of a stethoscope
17. Upper and lower eyelids separate and eyelashes form **18.** Eyes open
19. Developing baby can survive on its own if necessary **20.** Lung development becomes complete **Critical Thinking and Application 1.** Students may give the following reasons: Facts they already know about a developing baby; logical sequence of events. **2.** Answers will vary. **3.** Accept all logical questions.

Discovery Activity: Name It!
1. buck, doe, fawn **2.** drake, duck, duckling
3. bull, cow, calf **4.** dog, vixen, cub
5. gander, goose, gosling **6.** ram, ewe, lamb **7.** boar, sow, piglet **8.** bull, cow, calf **9.** cob, pen, cygnet **10.** buck, doe, kit **11.** buck, doe, joey **12.** billy, nanny, kid **13.** lion, lioness, cub **14.** bull, cow, pup **15.** bull, cow, calf

Activity: Stages of Human Development
1. a. 9 **b.** 2 **c.** 7 **d.** 6 **e.** 8 **f.** 3
g. 5 **h.** 4 **i.** 11 **j.** 1 **k.** 10 **2.** 9 months **3.** Check activity sheet.
4. Placenta **5.** Labor **6.** Check activity sheet.

Problem-Solving Activity: Male or Female?
1. Check karyotypes to make sure that they are correct. **2.** 23 (or 22 plus XY or XX)
3. Male **4.** There is one unmatched pair of chromosomes (XY).

Laboratory Investigation: How Many Offspring?
Observations 1. The horse. **2.** The opossum. **3.** The opossum. **4.** The human. **Analysis and Conclusions**
1. Generally, the longer the gestation period, the lower the number of offspring per birth. In some cases, the longer the life span, the longer the gestation period. **2.** 90 offspring. **3.** Humans, because they take the longest amount of time to reach puberty and need to be cared for longer. **4.** Students should use reference resources to identify needed data for the additional mammals snd then compare the information with their original data.

Contents

CHAPTER 8 ■ Immune System

Chapter Discovery
*Antigens and Antibodies................................H151

Chapter Activities
*Activity: Complete-a-Disease H155
*Activity: Call on the Experts in Infectious Diseases H157
Activity: AIDS...H159
Problem-Solving Activity: The Distribution of Disease H161
Problem-Solving Activity: Chronic Disorders H163
*Activity: Animal Disease Carriers H165

Laboratory Investigation Worksheet
Observing the Action of Alcohol on Microorganisms H167
(**Note:** *This investigation is found on page H210 of the student textbook.*)

Answer Key H169

*Appropriate for cooperative learning

Name _____ Class _____ Date _____

Chapter Discovery

CHAPTER 8 — Immune System

Antigens and Antibodies

Background Information

An antigen is any substance or organism that invades the body. Antibodies are proteins that are produced by certain types of white blood cells in response to an invader. Antibodies are part of the body's defense system against disease and infection.

Materials

2 sheets of colored construction paper
tracing paper
scissors
glue
piece of posterboard

Procedure

1. Using tracing paper, trace the shapes that appear at the end of this activity. Cut out each shape and then trace the shape onto the sheets of construction paper. Label each shape as shown.

2. Cut out the shapes from the construction paper. Then spread them out on a desk or table.

3. Find two shapes that fit together like pieces of a jigsaw puzzle. Glue the shapes in place on the piece of posterboard. (It does not matter where on the posterboard you place the shapes, as long as the shapes are joined together. Do keep in mind, though, that you eventually will have to fit all the shapes on the posterboard.)

4. Repeat step 3 until you have matched up all the shapes and glued them onto the posterboard.

Critical Thinking and Application

1. What relationship seems to exist between antigens and antibodies?

_____

_____

2. Is the relationship between antigens and antibodies specific or nonspecific? Explain.

_____

_____

_____

© Prentice-Hall, Inc.

Human Biology and Health H ■ 151

3. According to this model, what seems to happen when an antibody comes in contact with an antigen? *Hint:* Look at the shape of an antibody compared to that of an antigen.

_____

_____

_____

4. Why do you think the action of an antibody is helpful in preventing disease or infection?

_____

_____

_____

_____

_____

5. Sometimes it takes time for white blood cells to produce antibodies needed to attack an antigen. If this were the case in your body, what would you be experiencing while waiting for the antibodies to be produced?

_____

_____

6. Sometimes white blood cells "remember" how to make a certain type of antibody, and thus produce antibodies immediately. If this were the case in your body, what would happen if you were exposed to the disease that these antibodies were designed to fight?

_____

_____

Name _____ Class _____ Date _____

Human Biology and Health H ■ 153

Activity

Immune System

CHAPTER 8

Complete-a-Disease

Using the 26 letters of the alphabet, complete the words below. Use each letter only once. Cross out each letter after you use it. Each completed word will spell the name of an infectious disease. Beside each disease, indicate whether it is caused by a virus or a bacterium.

A B C D E F G H I J K L M N O P Q R S T U V W X Y Z

1. _ U M _ S _____

2. I _ _ L U _ N _ A _____

3. _ O C K _ A W _____

4. _ I P H _ H E R I A _____

5. _ C A _ L E T F E _ E R _____

6. P _ L _ O M _ E L I T I S _____

7. _ _ I C _ E N P O _ _____

8. _ U I N S Y _____

9. R _ _ I E S _____

10. _ H O O P I N _ C O _ G H _____

Name _____ Class _____ Date _____

Activity

Immune System

CHAPTER 8

Call on the Experts in Infectious Diseases

Hidden below are the names of twelve historical figures who made important contributions to the treatment of infectious diseases. To call on these experts, you need to decipher their names by using the letters on a telephone dial. Notice that a number can stand for more than one letter. You have to choose the right one! Clues 1 through 10 may help you. Match the clues with the names you have decoded.

A. 5 6 7 3 7 4   5 4 7 8 3 7
   _ _ _ _ _ _   _ _ _ _ _ _

B. 5 6 8 4 7   7 2 7 8 3 8 7
   _ _ _ _ _   _ _ _ _ _ _ _

C. 3 3 9 2 7 3   5 3 6 6 3 7
   _ _ _ _ _ _   _ _ _ _ _ _

D. 2 2 7 5   9 8 6 3 3 7 5 4 2 4
   _ _ _ _   _ _ _ _ _ _ _ _ _ _

E. 2 5 3 9 2 6 3 3 7   3 5 3 6 4 6 4
   _ _ _ _ _ _ _ _ _   _ _ _ _ _ _ _

F. 2 6 8 6 6   8 2 6   5 3 3 8 9 3 6 4 6 3 5
   _ _ _ _ _   _ _ _   _ _ _ _ _ _ _ _ _ _ _

G. 7 6 2 3 7 8   5 6 2 4
   _ _ _ _ _ _   _ _ _ _

H. 5 6 6 2 7   7 2 5 5
   _ _ _ _ _   _ _ _ _

I. 9 2 5 8 3 7   7 3 3 3
   _ _ _ _ _ _   _ _ _ _

J. 5 6 4 6   7 6 6 9
   _ _ _ _   _ _ _ _

Clues

1. Discovered penicillin _____
2. Proved that yellow fever is carried by mosquitoes _____
3. One of the first to draw sketches of bacteria he observed through a microscope he made _____
4. Isolated bacteria that caused such major diseases as anthrax and tuberculosis _____
5. Linked outbreaks of cholera to contaminated water _____
6. Developed vaccine for rabies and the germ-killing process called pasteurization _____
7. Performed the first vaccination against smallpox with material taken from a cowpox sore _____
8. Introduced the use of sterilized instruments, antiseptic dressings, and disinfectant spray in the operating room _____
9. Developed a vaccine for polio _____
10. Proved that fever was not a disease but a symptom of disease _____

Name _____ Class _____ Date _____

Activity

Immune System

CHAPTER 8

AIDS

Using reference materials in the library, find the following information about AIDS.

1. What does AIDS stand for? _____

2. What causes AIDS? _____

3. How does AIDS affect the body? _____

4. What are the symptoms of AIDS? _____

5. How is AIDS transmitted? _____

6. How is AIDS diagnosed? _____

7. Is there a treatment for AIDS? If so, describe it. _____

8. How can the transmission of AIDS be prevented? _____

© Prentice-Hall, Inc.

Human Biology and Health   H ■ 159

Name _____ Class _____ Date _____

Activity

Immune System

CHAPTER 8

The Distribution of Disease

These maps show how severely different kinds of diseases affect people living in different parts of the world. The darker the area, the more severe the disease. Study the maps and then answer the questions.

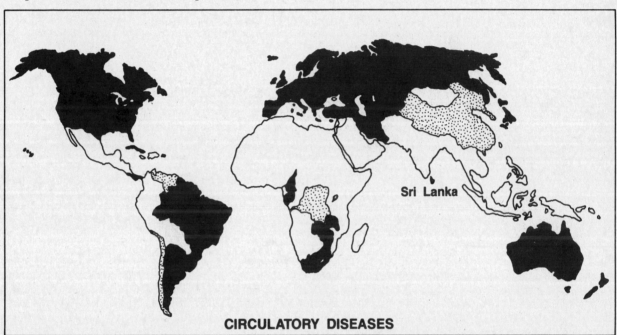

CIRCULATORY DISEASES

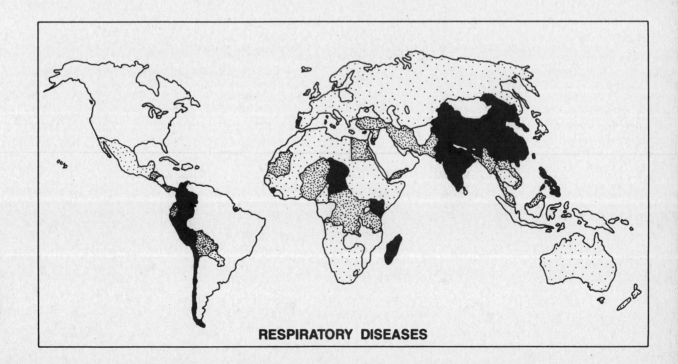

RESPIRATORY DISEASES

© Prentice-Hall, Inc.   Human Biology and Health   H ■ 161

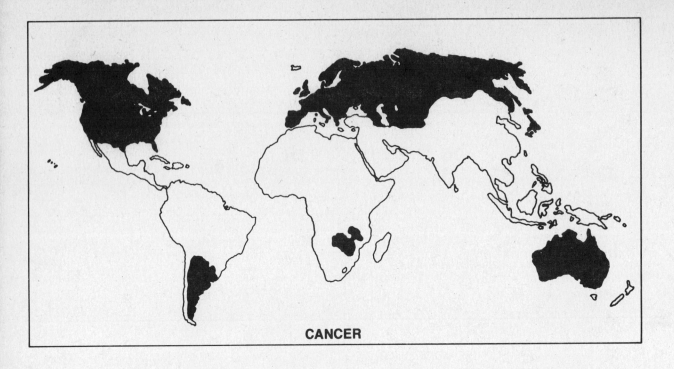

CANCER

1. In what country are circulatory diseases most severe?
   a. Mexico   b. India   c. United States   d. Egypt
2. On what continent are circulatory diseases least severe?
   a. North America   b. South America   c. Australia   d. Africa
3. In which country are respiratory diseases most severe?
   a. United States   b. Australia   c. South Africa   d. China
4. Where are respiratory diseases least severe?
   a. North America   b. Africa   c. Asia   d. Europe
5. In what country is cancer most severe?
   a. Japan   b. Egypt   c. India   d. Canada
6. On what continent is cancer least severe?
   a. North America   b. Australia   c. Africa   d. Europe
7. Notice that the island of Sri Lanka, only 32 km from India, has very different health statistics. How might you explain this difference? _____

   _____

8. In which country, India or Sri Lanka, are circulatory diseases a major killer? _____

9. What conclusions can you draw from these data? _____

Activity

Immune System

CHAPTER 8

Chronic Disorders

A. This circle graph shows the six leading causes of death in the United States. See if you can answer the questions by studying the information from the graph.

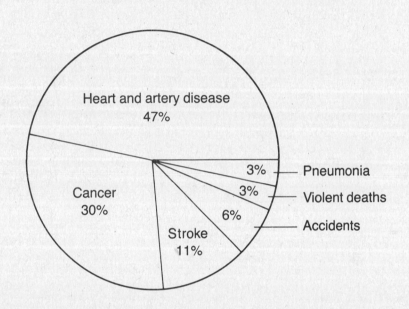

Six Leading Causes of Death in the United States

1. According to the graph, what is the leading cause of death in the United States? _____

2. About how many times as many Americans die from heart and artery disease as from cancer? _____

3. About how many times as many Americans die from cancer as from accidents? _____

4. About how many times as many people die from cancer as from violent deaths (homicides and suicides)? _____

5. If these six causes account for 1,500,000 deaths per year, approximately how many people die from heart and artery disease each year? _____
From cancer? _____ From pneumonia? _____

© Prentice-Hall, Inc.

Human Biology and Health

B. The following picture graph compares causes of death in the United States for the years 1970 and 1982. Answer the questions from the information given in the graph.

COMPARISON OF CAUSES OF DEATH IN THE UNITED STATES, 1970 AND 1982

	1970	1982
Chronic heart disease	👤👤👤👤👤👤👤👤👤👤👤👤👤👤👤 (15)	👤👤👤👤👤👤👤👤👤👤👤👤 (12)
Sudden heart attack	👤👤👤👤👤👤👤👤👤👤👤👤👤👤👤👤👤👤 (18)	👤👤👤👤👤👤👤👤👤👤👤👤👤👤👤👤 (16)
Lung cancer	👤👤👤▌ (3½)	👤👤👤👤👤 (5)
Breast cancer	👤👤 (2)	👤👤 (2)
Stroke	👤👤👤👤👤👤👤👤👤 (9)	👤👤👤👤👤👤 (6)
Auto accidents	👤👤▌ (2½)	👤👤 (2)
Pneumonia	👤👤👤 (3)	👤👤▌ (2½)

KEY: Each symbol represents 20,000 deaths

1. Which causes of death have decreased from 1970 to 1982? _____

   Which have increased? _____

2. Of those listed, which was the leading cause of death in both 1970 and 1982? _____

3. About how many times as many people die of sudden heart attacks as of pneumonia?

4. Give some possible reasons for a decline in chronic heart disease and heart attacks from 1970 to 1982. _____

164 ■ H Human Biology and Health

Name _____ Class _____ Date _____

Activity
Immune System
CHAPTER 8

Animal Disease Carriers

Drawings of animals that are carriers, or intermediate hosts, of infectious diseases are shown below. Can you identify the disease transmitted by each of these animals? Choose your answers from the following list of diseases.

African sleeping sickness	Bubonic plague	Chagas' disease
Rocky Mountain spotted fever	Rabies	Malaria
Schistosomiasis	Typhus	Yellow fever

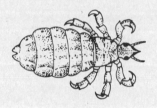

Human Body Louse

Tsetse Fly

1. _____

2. _____

Anopheles Mosquito

Aedes Mosquito

3. _____

4. _____

© Prentice-Hall, Inc.

Human Biology and Health   H ■ 165

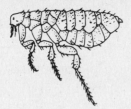

Rat Flea   Tick

5. _____   6. _____

Bat   Snail

7. _____   8. _____

Kissing Bug

9. _____

Name _____ Class _____ Date _____

Laboratory Investigation

CHAPTER 8 ■ Immune System

Observing the Action of Alcohol on Microorganisms

Problem
What effect does alcohol have on the growth of organisms?

Materials *(per group)*
glass-marking pencil
2 paper clips
2 thumbtacks
2 pennies
2 petri dishes with sterile nutrient agar
alcohol
100-mL beaker
transparent tape
graduated cylinder
forceps

Procedure 🧪 ☒

1. Obtain two petri dishes containing sterile nutrient agar.

2. Using a glass-marking pencil, label the lid of the first dish Soaked in Alcohol. Label the lid of the second dish Not Soaked in Alcohol. Write your name and today's date on each lid. **Note:** *Be sure to keep the dishes covered while labeling them.*

3. Using a graduated cylinder, carefully pour 50 mL of alcohol into a beaker.

4. Place a paper clip, a thumbtack, and a penny into the alcohol in the beaker. Keep these objects in the alcohol for 10 minutes.

5. Slightly raise the cover of the dish marked Not Soaked in Alcohol. **Note:** *Do not completely remove the cover from the dish.* Using clean forceps, place the other paper clip, thumbtack, and penny into the dish. Cover the dish immediately.

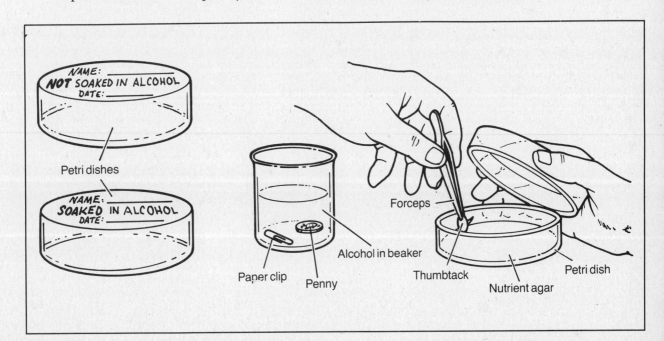

Human Biology and Health H ■ 167

6. Again using clean forceps, remove the paper clip, thumbtack, and penny from the alcohol in the beaker. Slightly raise the cover of the dish marked Soaked in Alcohol and place these objects into it.
7. Tape both dishes closed and put them in a place where they will remain undisturbed for 1 week.
8. After 1 week, examine the dishes. Make a sketch of what you see.
9. Follow your teacher's instructions for the proper disposal of all materials.

Observations
What did you observe in each dish after 1 week?
___
___

Analysis and Conclusions
1. What effect did alcohol have on the growth of organisms?
___
___

2. Why did you use forceps, rather than your fingers, to place the objects in the dishes?
___
___

3. Why did you have to close the petri dishes immediately after adding the objects?
___
___

4. Explain why doctors soak their instruments in alcohol.
___
___

Answer Key

CHAPTER 8 ■ Immune System

Chapter Discovery: Antigens and Antibodies
Critical Thinking and Application
1. Each antigen is paired with an antibody; they seem to fit together. **2.** Specific, because each antigen is paired with a particular antibody, and does not seem to fit with any other antibody. **3.** The antibody joins together with the antigen; the antibody seems to almost engulf or surround the antigen. **4.** By attaching itself to and partially surrounding the antigen, the antibody more or less immobilizes the antigen. It may slow the antigen down, keep it from moving, and/or prevent it from attaching to a cell. **5.** Symptoms of disease or infection—that is, you would be sick.
6. You would not get sick—you would be immune to the disease.

Activity: Complete-A-Disease
1. mumps—virus **2.** influenza—virus
3. lockjaw—bacterium **4.** diphtheria
—bacterium **5.** scarlet fever—bacterium
6. poliomyelitis—virus **7.** chicken pox
—virus **8.** quinsy—bacterium **9.** rabies
—virus **10.** whooping cough—bacterium
Quinsy is probably unfamiliar to most people. It is an acute throat infection that often occurs as a complication of tonsillitis.

Activity: Call on the Experts in Infectious Diseases
1. E—Alexander Fleming **2.** I—Walter Reed **3.** F—Anton van Leeuwenhoek
4. G—Robert Koch **5.** J—John Snow
6. B—Louis Pasteur **7.** C—Edward Jenner **8.** A—Joseph Lister **9.** H—Jonas Salk **10.** D—Carl Wunderlich

Activity: AIDS
1. AIDS (acquired immune deficiency syndrome) is a very serious disorder that causes severe damage to the body's defenses against disease. **2.** AIDS is caused by a virus named HIV (human immunodeficiency virus). **3.** HIV attacks helper-T cells, which are special white blood cells that are an important part of the body's immune system.
4. Some of the symptoms are similar to those of disorders that are less serious. However, with AIDS, the symptoms are usually prolonged. Some of the symptoms include enlarged lymph glands, fatigue, fever, loss of appetite and weight, and diarrhea. HIV may also infect the central nervous system. People with AIDS also suffer from other illnesses such as Kaposi's sarcoma, which is a form of cancer that usually arises in the skin and looks like a bruise, but grows and spreads.
5. AIDS is transmitted through sexual contact, exposure to infected blood, and from an infected pregnant woman to her fetus.
6. AIDS is diagnosed by means of a blood test. This test determines the presence of antibodies to HIV. **7.** An experimental drug called AZT (antiviral drug azidothymidene) has prolonged the lives of certain people with AIDS. But a cure for the disorder has not yet been found. **8.** To prevent transmission of HIV, the vector, or transmitter of the virus, must be avoided.

Problem-Solving Activity: The Distribution of Disease
1. c **2.** d **3.** d **4.** a **5.** d **6.** c
7. Different cultures, lifestyle, geography; perhaps a different gene pool **8.** Sri Lanka **9.** Environment plays an important role in some diseases.

Problem-Solving Activity: Chronic Disorders
A. 1. Heart and artery disease **2.** 1.5 times as many **3.** 5 times as many **4.** 10 times as many **5.** 705,000 people; 450,000 people; 45,000 people **B. 1.** Chronic heart disease, sudden heart attack, stroke, auto accidents, pneumonia; lung cancer, breast cancer

2. Sudden heart attack   3. 6 times as many   4. Greater public awareness of how to curb incidence of heart disease; better, more advanced methods of treatment

Activity: Animal Disease Carriers
1. Typhus   2. African sleeping sickness   3. Malaria   4. Yellow fever   5. Bubonic plague   6. Rocky Mountain spotted fever   7. Rabies   8. Schistosomiasis   9. Chagas' disease

Laboratory Investigation: Observing the Action of Alcohol on Microorganisms
Observations All dishes will contain circular colonies growing around the objects, but dishes marked Not Soaked in Alcohol will have far more colonies than dishes marked Soaked in Alcohol.   **Analysis and Conclusions**   1. It killed off or stopped the growth of most microorganisms.   2. Fingers may have transferred microorganisms from the hand to the dish.   3. To keep out any microorganisms from the air.   4. To kill certain microorganisms on their instruments that could cause infection in a patient.

Contents

CHAPTER 9 ■ Alcohol, Tobacco, and Drugs

Chapter Discovery
*Looking at Drugs H173

Chapter Activities
*Activity: Symptoms of Abuse H177
*Discovery Activity: Effects of Alcohol H179
Activity: Teenagers and Alcoholism H181

Laboratory Investigation Worksheet
Analyzing Smoking Advertisements H183
(**Note:** *This investigation is found on page H232 of the student textbook.*)

Answer Key H185

*Appropriate for cooperative learning

Name _____ Class _____ Date _____

Chapter Discovery Alcohol, Tobacco, and Drugs

CHAPTER 9

Looking at Drugs

Background Information
A drug is any substance that has an effect on the body. Drugs that are used to treat medical conditions are called medicines. Medicines can be classified as over-the-counter drugs or prescription drugs. Over-the-counter drugs are medicines that can be purchased without a doctor's prescription. Aspirin is an example of an over-the-counter drug. Prescription drugs can be obtained only with a doctor's prescription. Strong pain killers and antibiotics are examples of prescription drugs.

Materials
label from each of the following:
    over-the-counter drug
    prescription drug
    alcoholic beverage
    cola drink

Procedure
Examine the label from each of the above closely.

Observations
1. What information does the label from the over-the-counter drug contain?

   _____

   _____

   _____

   _____

2. Does the label from the over-the-counter drug contain any warnings? If so, what are they?

   _____

   _____

   _____

© Prentice-Hall, Inc. Human Biology and Health H ■ 173

3. What information does the label from the prescription drug contain?

_____

_____

_____

_____

4. Does the label from the prescription drug contain a warning? If so, what is it?

_____

_____

_____

_____

5. What information does the label from the alcoholic beverage contain?

_____

_____

_____

_____

6. Does the label from the alcoholic beverage contain a warning? If so, what does it say?

_____

_____

_____

_____

7. Does the warning indicate that drinking alcoholic beverages may be addictive?

_____

8. Based on what you know about alcoholism and the behavior of a person who has had too much to drink, do you think that this label accurately describes what may happen to a person who drinks? _____

9. Examine the label from a cola drink. Look at the list of ingredients. Do you see any ingredient listed that you would consider a drug? If so, what? _____

174 ■ H   Human Biology and Health

Name _____ Class _____ Date _____

10. Does this label contain any warnings? If so, what are they? _____
_____

Critical Thinking and Application

1. How does the label from an over-the-counter drug differ from the label from a prescription drug?
_____
_____
_____
_____

2. From the labels you have examined, do medicines change with time? If so, what do you think causes medicines to change?
_____
_____
_____
_____

3. Do you think that the labels on over-the-counter drugs give adequate information? If not, what do you think should be different?
_____
_____
_____
_____

4. How do you feel about the labels on prescription drugs?
_____
_____
_____

© Prentice-Hall, Inc.

Human Biology and Health

5. Why do you think that an alcoholic beverage label and a cola drink label were included in this activity?

___

___

___

___

6. Do you agree with the way alcoholic beverages are labeled? If not, what would you change?

___

___

___

___

7. Do you think that cola drink labels should carry warnings? If so, what do you think they should say?

___

___

___

___

Name _____ Class _____ Date _____

Activity

Alcohol, Tobacco, and Drugs

CHAPTER 9

Symptoms of Abuse

The column on the left lists seven substances that may be abused. The column on the right lists possible symptoms of such abuse. Write the letter of the correct symptom on the line before each substance.

Substance

____ 1. alcohol

____ 2. caffeine

____ 3. opiates

____ 4. tobacco

____ 5. marijuana

____ 6. tranquilizers

____ 7. LSD

Symptom

A. emphysema

B. hallucination

C. cirrhosis

D. depression

E. nervousness

F. time distortion

G. rapid physical dependence

© Prentice-Hall, Inc.

Human Biology and Health

Name _____ Class _____ Date _____

Activity

Alcohol, Tobacco, and Drugs

CHAPTER 9

Effects of Alcohol

Drinking too much alcohol can affect judgment and reaction time. Each year in the United States, over 95,000 deaths are blamed on drinking. Drivers who have been drinking are involved in 800,000 automobile accidents and 25,000 traffic deaths each year. What does drinking do to the body? In this activity you will simulate some of the effects of drunkenness without really drinking any alcohol. You will need to work with two partners during this activity.

Part A   Walk the Line

First, using chalk or masking tape, make a straight line about 3 meters long on the floor. One member of your group should walk from the beginning of this line to the end, putting the heel of one foot right against the toe of the other. A second member of the group should time how long it takes to walk from one end of the line to the other. The third group member should keep track of the number of times the walker accidentally misses the line.

To simulate getting drunk, the walker should spin around in place for 10 seconds. Be certain that the area is cleared of furniture and other obstacles. The two partners should stand nearby and act as "spotters" ready to catch the spinner if he or she starts to fall. **CAUTION:** *Do not be the spinner if you are under a doctor's care, have dizzy spells or heart problems, or are not able to participate in physical education classes.*

As soon as the person finishes 10 seconds of spinning, lead him or her to the beginning of the line and repeat the walk. Be sure to record the time and number of misses in the table provided. Repeat the procedure with the other team members.

Walk the Line

Name	Before		After	
	Time	Mistakes	Time	Mistakes

© Prentice-Hall, Inc.

Human Biology and Health

Part B Connect the Dots

The object of this test is to draw a wavy line through the dots as shown. Your partner will time you for 10 seconds while you connect as many dots as you can. At the end of the 10 seconds, give yourself one point for each dot you cross and one point each time you touch the top or bottom line without crossing over it.

Repeat the spinning procedure to get "drunk." Then repeat the dot test and see how many points you score this time. Record all answers in the table provided.

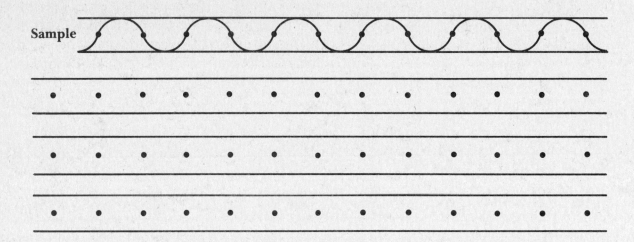

Connect the Dots

Name	Score Before	Score After

Name _____ Class _____ Date _____

Activity
Alcohol, Tobacco, and Drugs
CHAPTER 9

Teenagers and Alcoholism

A. It is estimated that there are 14 million alcoholics in the United States. Of this number, 3.5 million are teenagers, 13 to 17 years old. On the circle graph below, indicate the portion of the total population of alcoholics who are between the ages of 13 and 17. Among teenage alcoholics, 25 percent are females. Indicate this fact on the circle graph.

Alcoholics in the United States

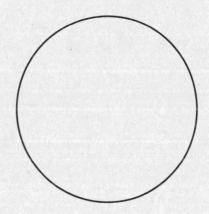

B. A recent survey of teenage drinking habits revealed the following.

   59 percent drank alcoholic beverages occasionally
   17 percent tried alcohol once
   23 percent never tried alcohol
   1 percent no comment

Graph the results of this survey in the circle below.

Teenage Drinking Habits

© Prentice-Hall, Inc.

Human Biology and Health   H ■ 181

Name _____ Class _____ Date _____

_____ Laboratory Investigation _____

CHAPTER 9 ■ Alcohol, Tobacco, and Drugs

Analyzing Smoking Advertisements

Problem
How are advertisements used to convince people to smoke or not to smoke?

Materials *(per group)*
magazines
paper

Procedure
1. Choose two or three different types of magazines. Glance through the magazines to find advertisements for and against cigarette smoking.
2. For each advertisement you found, fill in the information in the chart. In the last column, record the technique that the advertisement uses to attract the public to smoke or not to smoke. Examples of themes used to attract people to smoke are "Beautiful women smoke brand X"; "Successful people smoke brand Y"; "Brand Z tastes better." Examples of themes used to stop people from smoking are "Smoking is dangerous to your health"; "Smart people do not smoke"; "If you cared about yourself or your family, you would not smoke."

Magazine	Advertisements for Smoking (specify brand)	Advertisements Against Smoking (specify advertisement)	Theme

© Prentice-Hall, Inc.

Human Biology and Health H ■ 183

Observations

1. Were there more advertisements for or against smoking?

2. Which advertising themes were used most often? Least often?

Analysis and Conclusions

1. Which advertisements appealed to you personally? Why?

2. In general, how are the advertising themes that are used related to the type of magazine in which the advertisements appear?

3. **You and Your World** Repeat the procedure, but this time look for advertisements for and against drinking alcohol. Compare the way in which drinking alcohol is advertised to the way in which smoking cigarettes is advertised.

Answer Key

CHAPTER 9 ■ Alcohol, Tobacco, and Drugs

Chapter Discovery: Looking at Drugs
Observations 1. Ingredients; recommended dosage; information about use with children and/or people with special conditions; what the drug is for; expiration date; warnings of possible side-effects; what to do in case of overdose. **2.** Answers will vary depending on drug. **3.** Pharmacist's name; patient's name for whom the drug was prescribed; physician's name; date of prescription; expiration date; whether it can be refilled; warnings about side-effects; dosage; what the drug should be taken for; way the drug should be taken; what the drug contains. **4.** Answers will vary. **5.** Percentage of alcohol; volume of the bottle; warnings about use; may also include description of the beverage, such as the type of food a particular wine is meant to go with. **6.** Label should include a warning to pregnant women that drinking alcoholic beverages during pregnancy can cause birth defects; also that the use of alcohol impairs the ability to drive a car and operate machinery and may cause health problems. **7.** No. **8.** Accept all reasonable answers. **9.** Yes. Caffeine **10.** Cola drink label probably does not contain any warnings.

Critical Thinking and Application
1. Possible answers: over-the-counter drug contains advertising; prescription drug label is very specific, obviously typed for each container of medicine and addressed to a particular person who will use the drug; over-the-counter drug labels usually contain more detailed information about ingredients. **2.** Yes; the fact that both over-the-counter and prescription drug labels include expiration dates; change is caused by chemical reactions, such as the interaction of the drug with air and/or moisture. **3.** Accept all reasonable answers. **4.** Accept all reasonable answers. **5.** Both beverages contain substances that can be classified as drugs, because they have an effect on the body. These substances are alcohol and caffeine. **6.** Accept all reasonable answers. **7.** Accept all reasonable answers.

Activity: Symptoms of Abuse
1. C **2.** E **3.** G **4.** A **5.** F **6.** D **7.** B

Discovery Activity: Effects of Alcohol
Part A Answers will vary, but all students should have more mistakes after spinning.
Part B Answers will vary, but all students should have lower point scores after spinning.

Activity: Teenagers and Alcoholism
Graphs should accurately reflect the statistics given.

Laboratory Investigation: Analyzing Smoking Advertisements
Observations 1. Answers will vary, but generally speaking, a publication will contain more prosmoking than antismoking advertisements. **2.** Answers will vary, but in general, smoking advertisements depict attractive men and women engaged in sports or other forms of exercise. It is likely that the theme used least often is that smoking is harmful. **Analysis and Conclusions 1.** Answers will vary. Encourage students to analyze the reasons for their responses and the ways the advertisements seem to encourage them to smoke. **2.** In general, advertisements are designed to appeal to the type of person who reads the particular magazine in which the advertisement is found. **3.** Have students construct charts like the one they used for the smoking advertisements. After they have collected information about drinking advertisements, have them answer the same questions about the number of advertisements for and against, advertising themes used most and least often, personal appeal of advertisements, and relationship between advertising theme and type of magazine. Students will probably find that there are many similarities in the ways alcohol and cigarettes are advertised.

Science Reading Skills

TO THE TEACHER

One of the primary goals of the *Prentice Hall Science* program is to help students acquire skills that will improve their level of achievement in science. Increasing awareness of the thinking processes associated with communicating ideas and reading content materials for maximum understanding are two skills students need in order to handle a more demanding science curriculum. Teaching reading skills to junior high school students at successive grade levels will help ensure the mastery of science objectives. A review of teaching patterns in secondary science courses shows a new emphasis on developing concept skills rather than on accumulating factual information. The material presented in this section of the Activity Book serves as a vehicle for the simultaneous teaching of science reading skills and science content.

The activities in this section are designed to help students develop specific science reading skills. The skills are organized into three general areas: comprehension skills, study skills, and vocabulary skills. The Science Gazette at the end of the textbook provides the content material for learning and practicing these reading skills. Each Science Gazette article has at least one corresponding science reading skill exercise.

Contents

Adventures in Science
Claire Veronica Broome: Disease Detective
    Sequencing Events (Comprehension Skill) H191
    Vocabulary Skills (Vocabulary Skill) H191

Issues in Science
Are Americans Overexercising?
    Making Inferences (Comprehension Skill) H193
    Comparison-Contrast (Comprehension Skill) H194
    Evaluating Information (Study Skill) H195

Futures in Science
The Bionic Boy
    Using Context Clues (Vocabulary Skill) H197
    Reflective Writing (Study Skill) H199

Answer Key H201

Name _____ Class _____ Date _____

Science Gazette: Adventures in Science

Human Biology and Health

Claire Veronica Broome: Disease Detective
Science Reading Skill: Sequencing Events

The following is a list of events that are described in this article. After reading all the events, arrange them in the order in which they occurred. Write the letter of the event on the blank line next to the number that represents its sequence. In other words, put the letter of the event that occurred first on the blank next to the number one, and so on.

1. ____
2. ____
3. ____
4. ____
5. ____
6. ____
7. ____
8. ____

a. There was an epidemic of listeriosis in infants.
b. Dr. Broome and her team reviewed hospital records in Halifax.
c. Dr. Broome traced some contaminated coleslaw.
d. Dr. Broome returned to Atlanta.
e. Dr. Broome discovered that the bacteria that cause listeriosis had been identified in 1929.
f. She compared two groups: mothers of infants born with listeriosis and mothers who gave birth to healthy babies.
g. Dr. Broome noticed that women with sick babies had eaten more cheese than had women with healthy babies.
h. Dr. Broome went to Halifax.

Science Reading Skill: Vocabulary Skills

Having a large vocabulary is a key to understanding what you read. You can expand your vocabulary by determining the meaning of a word, remembering that meaning, and using the word as often as you can. Sometimes the meaning of a word is made clear by its context, or the way it is used in a sentence.

The following list of words was taken from this article. On the lines to the right of each word, write its definition. Then on the line to the left of the word, indicate how you determined its meaning. Use the letter "C" if you figured out the meaning from the context, "K" if you already knew the meaning, and "D" if you had to refer to the dictionary.

1. ____ apt _____

   _____

2. ____ membranes _____

   _____

© Prentice-Hall, Inc.

Human Biology and Health H ■ 191

3. _____ bacterium _____
   _____

4. _____ infect _____
   _____

5. _____ data _____
   _____

6. _____ transmitted _____
   _____

7. _____ epidemic _____
   _____

8. _____ contaminated _____
   _____

9. _____ environment _____
   _____

Name _____ Class _____ Date _____

Science Gazette: Issues in Science

Human Biology and Health

Are Americans Overexercising?
Science Reading Skill: Making Inferences

Inferences are conclusions or decisions that go beyond the given facts. You make inferences by reasoning with the facts. The ability to make inferences will help you pick up meanings that are not directly stated within the content. To make inferences, follow this simple approach.

1. Examine all the facts given in the content.
2. Look for clues to determine facts that are missing.
3. Use any previous knowledge or experience of your own.
4. Now combine all the information you have gathered and make an inference.
5. Remember, an inference must be based on valid facts.

Several statements are listed below. Some of these statements are directly stated in this article. Some represent inferences you can draw from this article. Others are unrelated to the article. Read each statement and determine which type it is. Use the following letter code to indicate your answer. Place the letter on the line that appears before the statement.

D: Direct Statement   I: Inference   U: Unrelated

_____ 1. Both Rory Bentley and James Fixx died of heart disorders.

_____ 2. There has been an increase in exercise-related injuries.

_____ 3. Continuous straining of muscles and tendons can cause bone damage in addition to severe pain.

_____ 4. Being guided in exercising properly is different from avoiding exercise.

_____ 5. Studies indicate that exercise is a national fad.

_____ 6. There are strong feelings about the benefits of exercising, especially running.

_____ 7. Soft surfaces are better for runners.

_____ 8. Calcium is an important substance for normal bone growth.

_____ 9. People have become more interested in exercising than in learning how to exercise.

_____ 10. There is a direct relationship between exercising and preventing heart attacks.

_____ 11. Some runners have acquired "eating disorders."

© Prentice-Hall, Inc.

_____ 12. Several studies indicate that regular exercise helps maintain overall good health.

_____ 13. James Fixx was a serious runner as well as an authority on running.

_____ 14. Age does not seem to be a factor among runners.

_____ 15. This article indicates that some exercise is good for you and no exercise is harmful.

Science Reading Skill: Comparison-Contrast

The way in which material is organized often provides clues about the relationship of ideas within paragraphs. Material that is organized to show how ideas are alike or different follows a comparison-contrast pattern. When you see this pattern, you should guide your reading to determine what topic or main idea is compared or contrasted. Read with the purpose of noting similarities and differences described by the details in the paragraphs.

Guide words such as "however," "in the same way," "but," "on the other hand," and "the difference between" are clues to help you identify the comparison-contrast pattern.

You can practice using the comparison-contrast pattern in reading information by completing the chart below. Read the instructions carefully before you write your answers in the spaces on the chart. Some of the benefits and disadvantages of exercising are discussed in this article. Main ideas about exercising are listed in column I. For each idea, write a brief sentence that supports it in column II. Then identify the contrasting view. In column III, support the contrasting view by writing a sentence that disagrees with the idea in column I.

Column I Ideas on Exercising	Column II Supports the Idea	Column III Disagrees With the Idea
Professionals in the medical field as well as other experts express some doubts about the benefits of exercise.		
Eating disorders have been linked to people who overexercise.		
The number of people exercising and the sales of "how to exercise" books have both increased.		
The risks of exercising are outweighed by the dangers of not exercising at all.		

194 ■ H   Human Biology and Health

Name _____ Class _____ Date _____

Science Reading Skill: Evaluating Information

Evaluating information requires that you analyze content and decide if the facts are valid. The ability to see the difference between facts and opinions will help you judge the reliability of what is stated in the material. Acquiring the skill of evaluating information will also help you become an independent learner.

Applying the skills you have just learned, write a brief paragraph that answers the question posed at the end of this article. Organize your ideas based on the facts in the article. After evaluating all the information, present your point of view. Use the space below to write your paragraph.

Name _____ Class _____ Date _____

Science Gazette: Futures in Science

Human Biology and Health

The Bionic Boy
Science Reading Skill: Using Context Clues

"Word power" is the ability to add to your science vocabulary. Because science terms are used to represent major ideas, acquiring the skill of defining words is important in reading and understanding science content. Using context clues is one of the best techniques for finding the meanings of words. There are five types of context clues you can use for defining science words.

1. **The Definition Clue.** The word is defined within the sentence structure or in a separate sentence. Example: *Synthesis* is a way in which new cells are formed. The process of combining small molecules to form larger ones is called synthesis.
   Synthesis means combining small molecules to form larger ones.

2. **The Synonym Clue.** The word is defined through a similar meaning of another word in the sentence. Example: Radio waves are sent in only one direction using a type of antenna that *focuses,* or beams, the waves in a particular direction.
   To focus means to beam the radio waves in a particular direction.

3. **The Comparison and Contrast Clue.** The word is defined through a meaning that is the same as or different from another word. Example: In any blood transfusion, it is important that the *donor* has the same blood type as the recipient, or person receiving the transfusion.
   Donor means the person giving blood.

4. **The Summary Clue.** The word is defined through connecting ideas that reinforce its meaning. Example: The *biome* of the tundra is so cold that part of the ground is frozen all year round. With winter temperatures well below the freezing point, this large region looks like a white sea of ice. Trees cannot grow in this area because they are unable to survive in this climate. The animals that live there have a coat of thick fur for protection against the cold.
   Biome means the climate conditions of a large region in which certain plants and animals live.

5. **The Experience Clue.** The word is defined by relating it to previous knowledge. Example: Early detection and treatment of cancer act as a *deterrent* to this fatal disease.
   Deterrent means a way of protection against cancer.

Learn how to use context clues to help you define words you do not know. As you develop vocabulary skills, you will also improve your comprehension of science concepts.

© Prentice-Hall, Inc.

A. A number of words have been selected from the content of this science article. Refer to the numbered paragraph to find each of the words. Find the definition of these words by using context clues. Circle the letter of the choice that best defines each word.

1. vaccines (paragraph 1)
   (a) used for prescriptions   (b) used to give protection against disease   (c) used to kill pain   (d) used to detect disease

2. physician (paragraph 2)
   (a) author   (b) accountant   (c) technician   (d) doctor

3. artificial (paragraph 4)
   (a) original   (b) not natural   (c) same   (d) different

4. transplant (paragraph 5)
   (a) remove organs   (b) reproduce organs   (c) replace organs   (d) transverse organs

5. indefinitely (paragraph 8)
   (a) originally   (b) partially   (c) without end   (d) quickly

6. trachea (paragraph 10)
   (a) liver   (b) lungs   (c) heart   (d) windpipe

7. capillaries (paragraph 11)
   (a) large blood vessels   (b) tiny blood vessels   (c) small tubes around the lungs   (d) air sacs

8. alveoli (paragraph 11)
   (a) air sacs of the lungs   (b) air sacs that carry blood   (c) large blood vessels   (d) tiny blood vessels

B. Identify the type of context clue you used to help you define each of the words. In some cases more than one clue may have been used. Write your answers on the lines provided.

1. vaccines _____

2. physician _____

3. artificial _____

4. transplant _____

5. indefinitely _____

6. trachea _____

7. capillaries _____

8. alveoli _____

198 ■ H   Human Biology and Health

Name _____ Class _____ Date _____

Science Reading Skill: Reflective Writing

Many of the skills used in reading science material are the same as those used in writing material that expresses your understanding of science concepts. The relationship between science reading and writing involves using the technical language of science to represent ideas. It requires a way of thinking that clearly communicates those ideas. Reflective writing is the ability to describe, relate, or summarize science content. The model below shows you the sequence of skills in reflective writing.

From Reading
* understand main ideas
* remember key words
* recall major details

Prewriting
* outline ideas
* connect ideas logically
* use words that express ideas of the science topic

Drafting
* select appropriate information
* organize ideas into sentences
* develop paragraphs to combine ideas

Rewriting in Final Form
* fulfill the purpose of the writing assignment
* check content for accurate information
* use proper punctuation and sentence formation

Based on the content of this science article, write three or four short paragraphs that tell what you learned from reading the letter sent by Dr. R. K. Smith to Evelyn. Refer to the model above to help you express your own ideas about this topic.

© Prentice-Hall, Inc.   Human Biology and Health   H ■ 199

Answer Key

Science Gazette

Adventures in Science
Sequencing Events
1. a 2. e 3. h 4. b 5. f 6. g 7. c
8. d **Vocabulary Skills** Methods used by students to determine meaning will vary. Possible answers: 1. likely 2. thin skin that lines an organ or body part 3. a one-celled organism 4. to make sick 5. available information 6. carried 7. rapidly spreading disease 8. exposed to disease-causing organisms 9. conditions and surroundings

Issues in Science
Making Inferences
1. D 2. D 3. I 4. I 5. U 6. D 7. I
8. I 9. I 10. D 11. D 12. D 13. I
14. I 15. I **Comparison-Contrast**
Column II Supports the Idea Doctors have seen an increase in exercise-related injuries. Overdoing athletic activity can stunt growth by damaging bones. **Column III Disagrees With the Idea** Studies indicate that regular exercise lowers the chances of heart attacks and reduces the likelihood of high blood pressure. **Column II Supports the Idea** Runners and others who exercise tend to undereat because they have a fear of losing their athletic look. **Column III Disagrees With the Idea** People should learn to eat the proper foods and to exercise under a doctor's care. **Column II Supports the Idea** People are using books, records, videocassettes, and TV programs to learn how to exercise. **Column III Disagrees With the Idea** Health experts say that a prescription for exercise is needed. **Column II Supports the Idea** More Americans die from sitting around than from running around. **Column III Disagrees With the Idea** People are too careless about exercising properly and can injure themselves. **Evaluating Information** Answers will vary.

Futures in Science
Using Context Clues
A. 1. b 2. d 3. b 4. c 5. c 6. d 7. b
8. a B. 1. Experience clue 2. Definition clue and/or Summary clue 3. Comparison and contrast clue 4. Summary clue
5. Comparison and contrast clue
6. Synonym clue 7. Synonym clue 8. Synonym clue **Reflective Writing** The content of paragraphs vary according to the way students decide to present their ideas.

Activity Bank

татко THE TEACHER

One of the most exciting and enjoyable ways for students to learn science is for them to experience it firsthand—to be active participants in the investigative process. Throughout the *Prentice Hall Science* program, ample opportunity has been provided for hands-on, discovery learning. With the inclusion of the Activity Bank in this Activity Book, students have additional opportunities to hypothesize, experiment, observe, analyze, conclude, and apply—all in a nonthreatening setting using a variety of easily obtainable materials.

These highly visual activities have been designed to meet a number of common classroom situations. They accommodate a wide range of student abilities and interests. They reinforce and extend a variety of science skills and encourage problem solving, critical thinking, and discovery learning. The required materials make the activities easy to use in the classroom or at home. The design and simplicity of the activities make them particularly appropriate for ESL students. And finally, the format lends itself to use in cooperative-learning settings. Indeed, many of the activities identify a cooperative-learning strategy.

Students will find the activities that follow exciting, interesting, entertaining, and relevant to the science concepts being learned and to their daily lives. They will find themselves detectives, observing and exploring a range of scientific phenomena. As they sort through information in search of answers, they will be reminded to keep an open mind, ask lots of questions, and most importantly, have fun learning science.

Contents

Chapter 1 The Human Body
A Human Cell vs. an Ameba..........................H207

Chapter 2 Skeletal and Muscular Systems
Under Tension......................................H211

Chapter 3 Digestive System
Getting the Iron Out...............................H213
Going Crackers.....................................H217
Do You Have the Stomach for ...?...................H221

Chapter 4 Circulatory System
A Pulsating Question...............................H225
You've Got to Have Heart...........................H229
The Squeeze Is On..................................H231

Chapter 5 Respiratory and Excretory Systems
Taking a Breather..................................H235
Lip Service?.......................................H239
How Fast Do Your Nails Grow?.......................H243

Chapter 6 Nervous and Endocrine Systems
How Fast Can You React?............................H245
A Gentle Touch.....................................H249
Colored "Sandwiches"...............................H253
More Than Meets the Eye............................H257
Some Sound Reasons.................................H259

Chapter 8 Immune System
It's No Skin Off Your Nose.........................H263

Answer Key.....................................H267

Name _____ Class _____ Date _____

Activity
The Human Body — CHAPTER 1

A Human Cell vs. an Ameba

What do you and a single-celled ameba have in common? Perhaps, you may think nothing. But if you were to look a little closer, you would see that you and an ameba have a lot more in common than you thought. Why not try this activity and find out for yourself.

Materials
flat toothpick
2 microscope slides
2 medicine droppers
2 coverslips
methylene blue
paper towel
microscope
ameba culture
pencil

Procedure

1. Put a drop of water in the center of a microscope slide.
2. Using the end of the toothpick, gently scrape the inside of your cheek. Even though you cannot see them, there will be cheek cells sticking to the toothpick.
3. Stir the scrapings into the drop of water on the slide.
4. To make a wet-mount slide, use the tip of the pencil to gently lower the coverslip over the cheek cells.

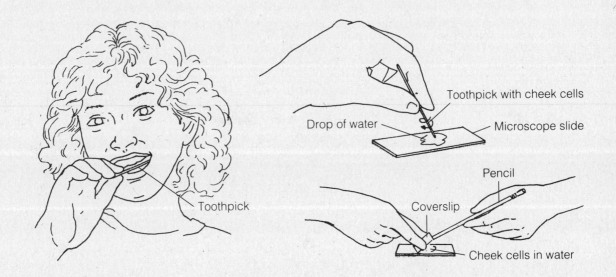

© Prentice-Hall, Inc.

Human Biology and Health H ■ 207

5. With a medicine dropper, put one drop of methylene blue at the edge of the coverslip. **CAUTION:** *Be careful when using methylene blue because it may stain the skin and clothing.*
6. Place a small piece of paper towel near the edge of the coverslip and allow the paper towel to absorb the excess methylene blue.

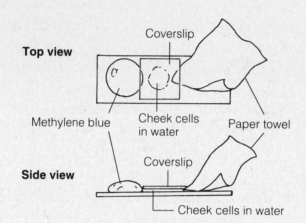

7. With a medicine dropper, have your partner place a drop of the ameba culture on the other microscope slide.
8. Your partner should make a wet-mount slide of the ameba culture.

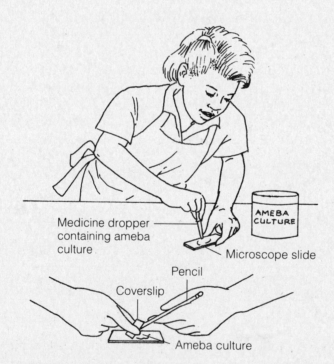

9. To remove any excess liquid, your partner should repeat step 6.
10. Place the slide of your cheek cells on the stage of the microscope and locate a cheek cell under low power. Then switch to the high-power objective lens.
11. Observe a cheek cell and sketch what you see. Label the cell parts that you see.
12. Have your partner repeat steps 10 and 11 using the slide of the ameba culture.

Name _____ Class _____ Date _____

13. Reverse roles with your partner and repeat steps 1 through 12.
14. Complete the Data Table below. If the cheek cell or the ameba contains any of the structures listed in the Data Table, write the word present in the appropriate place. If the cheek cell or the ameba does not have the structure, write the word absent.

Observations
DATA TABLE

Cell	Cell Membrane	Nucleus	Other
Human cheek			
Ameba			

Analysis and Conclusions

1. How are a human cheek cell and an ameba similar?

   _____
   _____
   _____
   _____

2. How are they different?

   _____
   _____
   _____
   _____

3. How many cells are you made of? An ameba?

   _____
   _____

4. Share your results with those of your classmates. Did the findings of any of your classmates differ from yours? Can you explain why there was a difference?

   _____
   _____
   _____

© Prentice-Hall, Inc.   Human Biology and Health

Activity

Skeletal and Muscular Systems

CHAPTER 2

Under Tension

Muscles are the only body tissues that are able to contract, or tighten up. You coordinate the movements of muscles without thinking by relaxing one muscle while you tighten another. Some movements, such as holding your hand out in front of you, are so slight that they go unnoticed. However, you can make these movements visible by doing this activity. All you will need is a table knife and a hairpin.

What You Will Do
1. Have your partner hold the knife out in front of him or her parallel to the top of a table or level desk. **Note:** *Your partner should not touch the table with his or her hand or arm.*
2. Place the hairpin on the knife as shown in the bottom diagram.

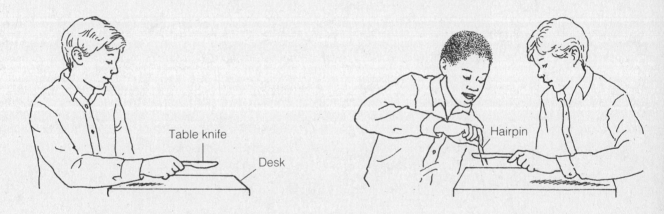

3. Have your partner raise the knife just high enough off the table for the "legs" of the hairpin to touch the table and the "head" of the hairpin to rest on the edge of the knife.

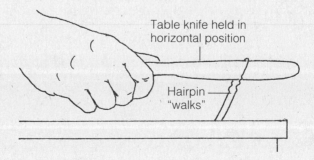

4. Have your partner hold the knife as steady as he or she can for 20 seconds. Observe what happens to the hairpin.
5. Repeat step 4, having your partner tighten his or her hold on the knife. Observe what happens.
6. Reverse roles with your partner and repeat steps 1 through 5. Share your results with your classmates.

What You Will Discover
1. What happened to the hairpin in step 4? What does this action show?

   _____
   _____
   _____

2. What happened to the hairpin when the hold on it was tightened? How do you explain this action?

   _____
   _____
   _____
   _____
   _____

3. Did your classmates have similar results?

   _____

Going Further

Repeat the activity but this time increase the time given in steps 4 and 5 to 1 minute. Observe what happens to the hairpin. How can you relate this activity to what happens when you have to stand in the same place for a long time?

Name _____ Class _____ Date _____

Activity

Digestive System

CHAPTER 3

Getting the Iron Out

Have you ever read the list of ingredients on the side panel of your box of breakfast cereal? Perhaps you should. The list of ingredients contains some important information about the nutrients your body needs. Nutrients include carbohydrates, proteins, fats, minerals, and vitamins. Sometimes nutrients are added to breakfast cereals, forming mixtures (combinations of substances that are not chemically combined). How can you separate one substance from another in a mixture? Try this activity to find out how you can separate the mineral iron from a breakfast cereal.

Materials
balance
breakfast cereal with 100 percent
　of the recommended daily
　requirement (RDA) of iron
sealable plastic bag
large plastic container

bar magnet
clock or timer
white tissue paper
hand lens
spoon

Procedure
1. Measure out 50 g of a breakfast cereal. Place the cereal into a sealable plastic bag, pressing down on it to remove most of the air inside. **CAUTION:** *Do not eat any foods or drink any liquids during this activity.*

© Prentice-Hall, Inc.

Human Biology and Health　H ■ 213

2. With your hands, crush the cereal into a fine powder and pour it into the plastic container. Add enough water to completely cover the cereal.

3. Use a bar magnet to stir the cereal-and-water mixture for at least 10 minutes.

4. Remove the bar magnet from the mixture. Allow any liquid on the magnet to drain off.

5. Use a piece of tissue paper to scrape off any particles that are attached to the bar magnet. Observe the particles with a hand lens.

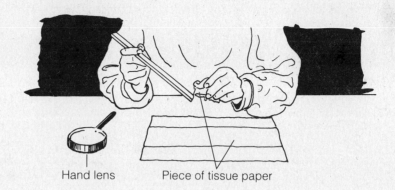

Observations

What did you observe when you looked at the particles scraped from the sides of the bar magnet?

_____
_____
_____
_____

Analysis and Conclusions

1. How do you know that the breakfast cereal is a mixture?

_____
_____
_____
_____

2. What do you think the particles scraped from the bar magnet are?

_____
_____
_____

3. Why do you think it was possible to separate the particles from the rest of the cereal?

_____
_____
_____
_____
_____

4. Do you think the cereal is a mixture made up of different materials unevenly spread out or a mixture of materials uniformly spread out? Explain.

_____
_____
_____
_____
_____

Going Further

Repeat the procedure with 50 g of another cereal that contains less than 100 percent of the RDA for iron. Can you separate the iron from this cereal?

Name _____ Class _____ Date _____

Activity
Digestive System CHAPTER 3

Going Crackers

As you might guess from the word itself, carbohydrates are made of carbon (*carbo-*), hydrogen (*hydr-*), and oxygen (*-ate*). Foods that are rich in carbohydrates are the ones that contain starches and sugars. Your body can perform a little "magic" and change the starches into a sugar (called glucose), which you need for energy. In the mouth, a substance in saliva digests (breaks down) starch into sugar. The substance is ptyalin. The following activity will show you how quickly this change takes place in your mouth.

Materials
glass-marking pencil   drinking cup
4 test tubes   large beaker
test-tube rack   graduated cylinder
white soda cracker   Benedict's solution
iodine solution   Bunsen burner
medicine dropper   tripod
spoon   wire gauze

Procedure
1. With a glass marking pencil, label the test tubes from 1 to 4.
2. Divide a white soda cracker in half. Put one half of the soda cracker aside for use in step 9.
3. Divide the other half of the soda cracker into two equal pieces. Place one piece into test tube 1 and the other piece into test tube 2.
4. Add 3 drops of iodine solution to test tube 1. Iodine solution will turn black in the presence of a starch.

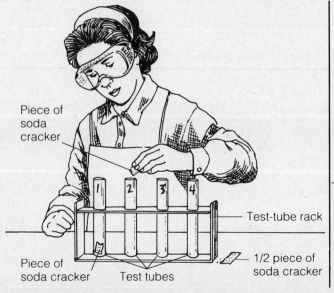
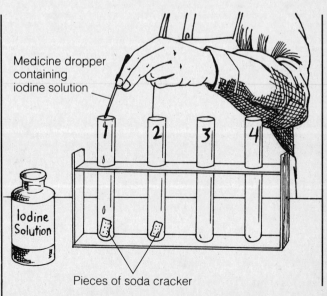

Human Biology and Health H ■ 217

5. Record your observations in the Data Table.
6. Add 5 mL of Benedict's solution to test tube 2.
7. Place the test tube in a beaker of boiling water for a few minutes. **CAUTION:** *Be careful when heating substances.* If the mixture changes color to green or reddish orange, sugar is present.

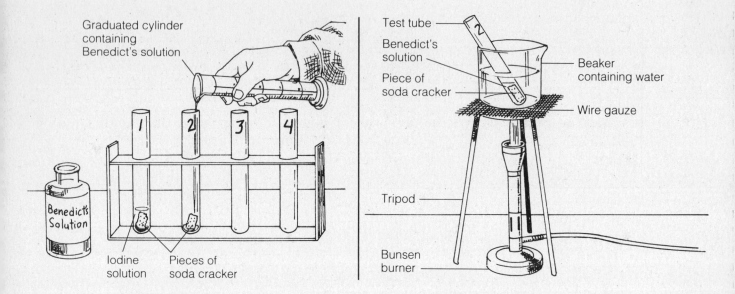

8. Record your observations in the Data Table.
9. Rinse your mouth out with water and chew on the remaining half of the soda cracker for 5 minutes.
10. After 5 minutes, use a spoon to transfer half of the chewed soda cracker from your mouth into test tube 3 and the other half into test tube 4.
11. Repeat steps 4 and 5 using test tube 3.
12. Repeat steps 6 through 8 using test tube 4.

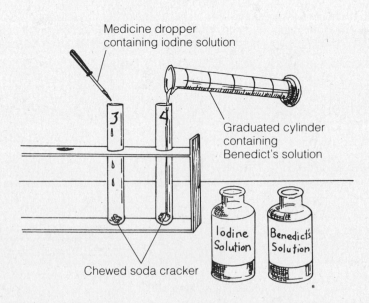

Name _____ Class _____ Date _____

Observations
DATA TABLE

Test Tube	Test for Starch or Sugar	Present or Absent
1 (Unchewed cracker, iodine solution)		
2 (Unchewed cracker, Benedict's solution)		
3 (Chewed cracker, iodine solution)		
4 (Chewed cracker, Benedict's solution)		

Analysis and Conclusions

1. Why was iodine solution added to test tubes 1 and 3?

   _____
   _____
   _____
   _____

2. Why was Benedict's solution added to test tubes 2 and 4? _____
   _____
   _____

3. How did the soda cracker taste when you began chewing? After 5 minutes?
   _____
   _____

4. What happened to the starch in the soda cracker after you chewed it? _____
   _____
   _____

© Prentice-Hall, Inc.   Human Biology and Health

5. What do the results of this activity tell you about what happens to starch in your mouth? _____

_____

_____

6. Compare your results with those of your classmates. Were they similar? Different? Explain. _____

_____

_____

_____

_____

_____

Going Further

You may wish to find out if temperature can affect the rate at which the reaction in this activity takes place. To do this, place an entire white soda cracker in your mouth and chew it a few times. Then divide it equally among 3 test tubes. Place one test tube in cold water, the second in hot water, and the third in water that is at room temperature. Then test the contents of the test tubes for sugar by repeating steps 6 and 7 in the activity.

Name _____ Class _____ Date _____

Activity

Digestive System

CHAPTER 3

Do You Have the Stomach for . . . ?

Just as the digestion of carbohydrates begins in the mouth, the digestion of proteins begins in the stomach. The substances that digest proteins, however, are different from the ones that digest carbohydrates. To see what substances are needed to digest a protein (egg white), why not try doing this activity.

Materials
3 test tubes
test-tube rack
glass-marking pencil
scalpel
boiled egg white
10-mL graduated cylinder
3 stoppers
pepsin
dilute hydrochloric acid
glass stirring rod
litmus paper

Procedure

1. Label 3 test tubes A, B, and C and place them in a test-tube rack.
2. With a scalpel, carefully cut the boiled egg white (protein) into nine 5-mm cubes.
3. Add 3 cubes of boiled egg white to each test tube.

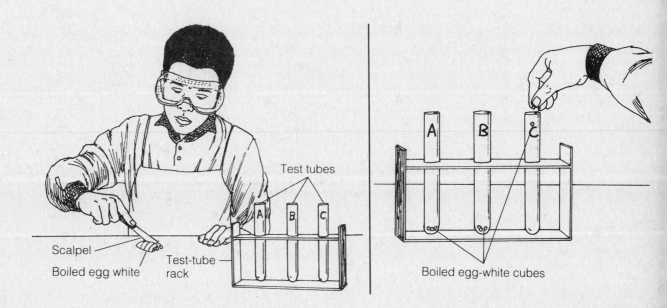

© Prentice-Hall, Inc.

Human Biology and Health   H ■ 221

4. Then add the following to each test tube. **CAUTION:** *Be careful when using an acid because it may burn your skin.*

   Test tube A—10 mL of pepsin
   Test tube B— 5 mL of pepsin and 5 mL of hydrochloric acid
   Test tube C—5 mL of pepsin and 5 mL of water

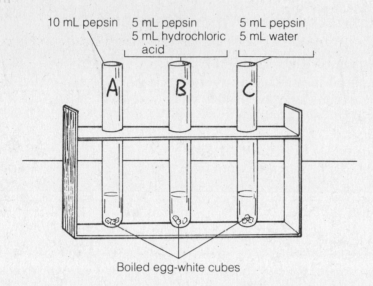

5. Examine each test tube and note the size and appearance of the pieces of egg white. Record your observations in the appropriate places in Data Table 1.
6. Using the stirring rod, pick up some of the contents of test tube A. Touch the stirring rod to a piece of blue litmus paper. Blue litmus paper will turn pink in the presence of an acid. Record the results in Data Table 2.

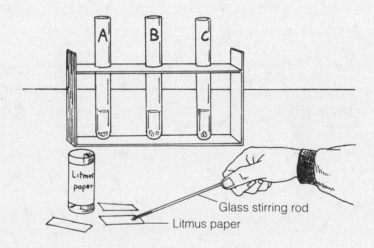

7. Repeat step 6 for test tubes B and C.

222 ■ H   Human Biology and Health

Name _____ Class _____ Date _____

8. Stopper the three test tubes and put them in a place where they will remain undisturbed for one day.
9. After one day, examine the test tubes and note any changes in the size and appearance of the pieces of egg whites. Record your observations in the appropriate data table.
10. Test the contents of each test tube with litmus paper again and record the results in the appropriate data table.

Observations
DATA TABLE 1

Test Tube	Egg White Size and Appearance at Start	Egg White Size and Appearance After One Day
A (egg white, pepsin)		
B (egg white, pepsin, hydrochloric acid)		
C (egg white, pepsin, water)		

DATA TABLE 2

Test Tube	Litmus Color and Appearance at Start	Litmus Color and Appearance After One Day
A (egg white, pepsin)		
B (egg white, pepsin, hydrochloric acid)		
C (egg white, pepsin, water)		

© Prentice-Hall, Inc.

Human Biology and Health   H ■ 223

Analysis and Conclusions

1. What effect did the addition of the hydrochloric acid have on the digestion of the egg white? _____
_____
_____

2. Why did you have to set the test tubes aside for one day? _____
_____
_____
_____

3. Which materials were the best at digesting the egg white? The worst? How were you able to tell? _____
_____
_____
_____
_____

4. What does this activity tell you about the digestion of protein in the stomach?
_____
_____
_____
_____

5. Share your results with your classmates. Did they have the same results as you did? Explain your answer. _____
_____
_____
_____
_____
_____

Name _____ Class _____ Date _____

Activity

Circulatory System

CHAPTER 4

A Pulsating Question

What do you and an earthworm have in common? At first glance, you may think nothing. However, if you take a closer look, you may discover an interesting similarity. What is it? Why not try this activity and find out. All you will need is a culture dish, a paper towel, a live earthworm, water, a hand lens, and a clock with a second indicator.

What Do You Do?
1. Select one member of each group for the following roles: Principal Investigator, Timer, and Observer/Recorder. Make sure you understand your role in the activity before you continue.
2. Line a culture dish with a moist paper towel and carefully place a live earthworm in it. Be sure to keep the earthworm moist by placing a few drops of water on it from time to time.
3. Using a hand lens, look for the blood vessel along the top surface of the earthworm. Watch the skin above the blood vessel until you see it move up and down. The up-and-down movement is the earthworm's pulse.

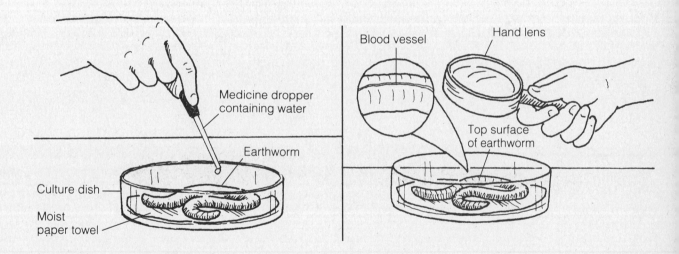

4. Count the number of times the earthworm's pulse beats in one minute. Record this number in the Data Table on the next page.
5. Repeat step 3 two more times, recording the earthworm's number of pulse beats per minute in the Data Table. Then find out the earthworm's average pulse for one minute.

© Prentice-Hall, Inc.

Human Biology and Health   H ■ 225

6. Now, count the number of times your pulse beats in one minute. To locate your pulse, place the index and middle finger of one hand on the wrist of your other hand where it joins the base of your thumb. Move the two fingers slightly until you locate your pulse. Record the number of times your pulse beats in one minute in the Data Table.

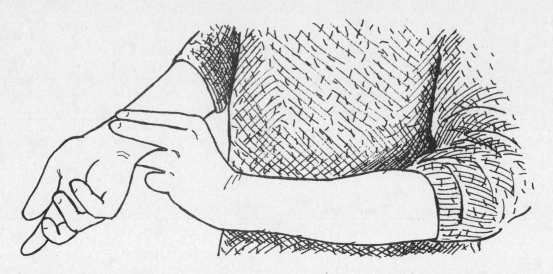

7. Repeat step 6 two more times, recording the number of times your pulse beats per minute in the Data Table. Then find out your average pulse for one minute and record it in the Data Table.

What Did You See?

DATA TABLE

Trial	Pulse Rate (per minute)	
	Earthworm	You
1		
2		
3		
Average		

What Did You Discover?

1. Why was it important to measure the earthworm's pulse and your pulse several times and then take the average of each? _____
_____
_____
_____
_____

Name _____ Class _____ Date _____

2. How does the earthworm's pulse rate compare with yours?
_____
_____

3. How are you and the earthworm similar? _____
_____
_____
_____

4. How do your results compare with those of your classmates? If there were any differences, can you explain why? _____
_____
_____
_____
_____
_____

Going Further
Design an activity in which you would show how exercise can affect your pulse rate.

Name _____ Class _____ Date _____

Activity

Circulatory System

CHAPTER 4

You've Got to Have Heart

Just as the mechanical pump in the water system in your home provides a constant pressure to force fluids along the pipes when you open the tap, so too does your heart provide pumping pressure. The job of any pump is to produce pressure and move a certain amount of fluid in a specific direction at an acceptable speed. Fluids, like blood, travel from an area of high pressure to an area of lower pressure.

Try to simulate the pumping action of your heart by doing this activity. You will need two plastic bottles of the same size, two one-hole rubber stoppers, and two 15-cm plastic tubes.

What You Will Do
1. Fill the two plastic bottles with water.
2. Insert the plastic tubes into the one-hole rubber stoppers. Then insert the rubber stoppers containing the plastic tubes into each bottle.

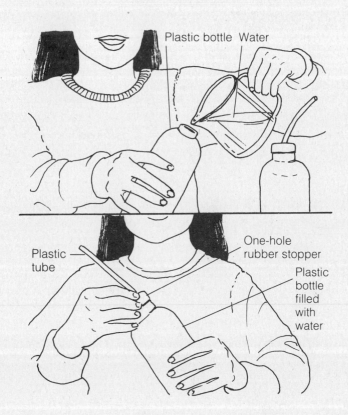

© Prentice-Hall, Inc.

Human Biology and Health H ■ 229

3. Have your partner squeeze one bottle with one hand, while you squeeze the other bottle with two hands. Observe what happens.

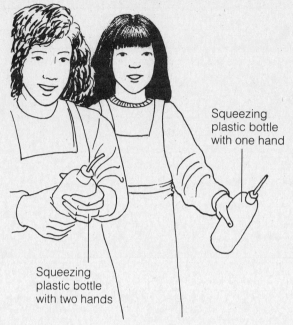

What You Will Discover
1. Which action squirted water further? Explain. _____
_____
_____
_____

2. Which action better simulated the pumping action of the heart?
_____
_____

3. What general statement can you make about pressure and the distance water travels?
_____
_____

4. When blood is pumped out of the heart, where does it go?
_____

5. How do your results and your partner's results compare with the class's results? Are they similar? Are they different? Explain why. _____
_____
_____
_____

Name _____ Class _____ Date _____

Activity

Circulatory System

CHAPTER 4

The Squeeze Is On

If you could look inside an artery and a vein, you would discover that the wall of the artery is thicker and more muscular than the wall of the vein. Might this characteristic make the blood pressure in arteries higher than the blood pressure in veins? Let's see.

What Do You Need?
plastic squeeze bottle
15-cm plastic tube
two-hole rubber stopper
15-cm glass tube
2 large desk blotters
meterstick
scissors
transparent tape

What Do You Do?
1. Fill a plastic squeeze bottle with water.

2. Place a plastic tube into one of the holes in a two-hole rubber stopper. Place a glass tube of the same length into the second hole of the stopper.

3. Insert the rubber stopper into the plastic bottle.

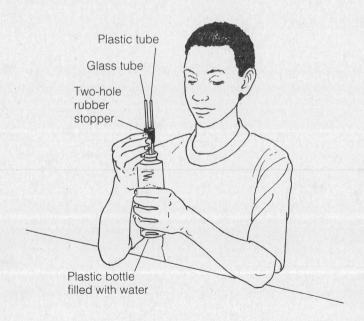

© Prentice-Hall, Inc.

Human Biology and Health H ■ 231

4. Cut two large desk blotters into 30-cm strips. Tape several strips together to form three 120-cm strips.

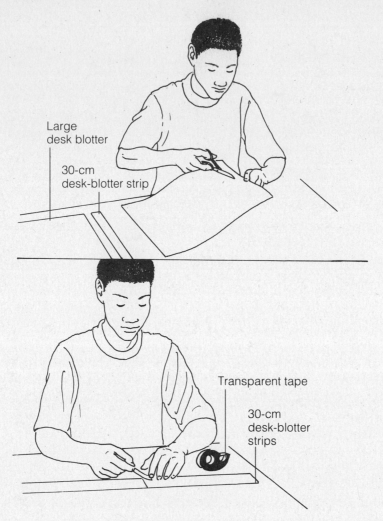

5. Place one of the blotter strips on the floor. Kneel on the floor with your knees touching one end of the blotter strip.
6. Hold the plastic bottle so that the glass tube is on the left and the plastic tube is on the right. Firmly squeeze the plastic bottle once.
7. With a meterstick, measure the distance from the edge of the blotter to where the wet spot for each tube first appears on the blotter. This is the distance water travels from each tube. Record this information in the Data Table on the next page.

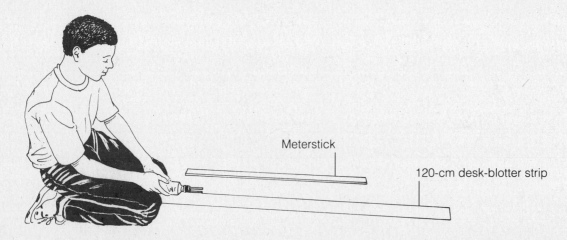

Name _____ Class _____ Date _____

8. Repeat steps 5 through 7 two more times using the remaining blotter strips.
9. Calculate the average distance the water travels from each tube. Record the average distance for each tube.

What Did You See?

DATA TABLE

Trial	Distance Water Traveled (cm)	
	Glass Tube	Plastic Tube
1		
2		
3		
Average		

Which tube squirted water farther? _____

What Did You Discover?

1. Which tube represents an artery? A vein? _____
   _____
   _____

2. What does the water in the tubes represent? _____

3. What does this activity tell you about blood pressure in arteries and veins?
   _____
   _____
   _____
   _____

4. What is the name of the body structure that forces the blood through the blood vessels? _____

5. How do your results compare with those of your classmates? If there were any differences, can you explain why? _____
   _____

© Prentice-Hall, Inc.   Human Biology and Health   H ■ 233

Name _____ Class _____ Date _____

Activity

Respiratory and Excretory Systems

CHAPTER 5

Taking a Breather

What do playing soccer, walking up a flight of stairs, and chasing after your younger sister or brother have in common? If your answer is that they are all forms of exercise, you are correct. If you do these activities regularly, you firm up your muscles, increase your endurance, and develop greater strength. In addition, regular exercise makes the heart strong enough to pump more blood with every beat. The same is true for the lungs. With each breath, strong lungs take in more oxygen and rid the body of more carbon dioxide. Let's do this activity and see if the lungs do work harder when exercising.

What Do You Need?
graduated cylinder
limewater
2 test tubes
clock with second indicator
2 drinking straws

What Must You Do?

1. Pour 5 mL of limewater into one test tube.
2. Place a straw in the test tube.
3. Have your partner begin timing you as you gently exhale through the straw into the limewater. Inhale through your nose. **CAUTION:** *Do not inhale through the straw.* Begin counting the number of breaths that you take.

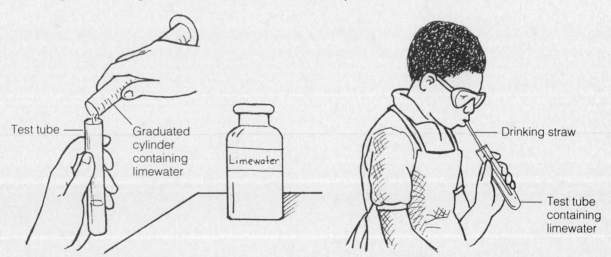

4. Continue inhaling through your nose and exhaling through your mouth until the limewater turns cloudy. This reaction is a test for carbon dioxide.

© Prentice-Hall, Inc.

Human Biology and Health

5. In the Data Table, record the time and the number of breaths it took for the limewater to turn cloudy.
6. Repeat steps 1 and 2 using a clean test tube and straw.
7. Run in place for 2 minutes. **CAUTION:** *If your doctor prohibits this type of activity, do not do this part.*
8. Repeat steps 3 and 4.

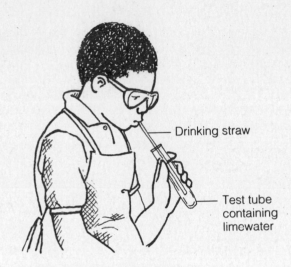

9. Using 2 new straws, reverse roles with your partner and repeat steps 1 through 8.

What Did You See?

DATA TABLE

Activity	Number of Minutes	Number of Breaths
At rest		
After exercise		

What Did You Discover?

1. Does limewater turn cloudy faster while resting or after exercise? _____

2. Do the lungs work harder while exercising? Give evidence to support your answer.

236 ■ H Human Biology and Health

Name _____ Class _____ Date _____

3. What is the relationship between the amount of carbon dioxide in your breath before and after exercise? _____
_____
_____
_____

4. Compare your results with those of your partner and your classmates. Explain any differences. _____
_____
_____
_____
_____

5. Suppose you are planning to run in a race that takes place at an altitude that is higher than what you are used to. What effects would the higher altitude have on your breathing? (*Hint:* There is less oxygen at higher altitudes.) _____
_____
_____
_____
_____
_____

Name _____ Class _____ Date _____

Activity
Respiratory and Excretory Systems

CHAPTER 5

Lip Service?

Chapped lips are not fun to have. Why do your lips dry out faster than the rest of the skin on your face? Unlike the rest of your face, the lips are covered only by a thin layer of skin. As a result, water is lost faster from the lips than it is from the rest of the face. What can you do to prevent your lips from drying out? Try this activity.

Materials
scissors
white tissue paper
petroleum jelly
metric ruler
white paper towel
cardboard
food coloring
plastic container
spoon
transparent tape
medicine dropper

Procedure
1. Cut two 4-cm × 7-cm strips of white tissue paper. Apply petroleum jelly to one strip and gently rub it in with your finger, making sure you have covered the middle of the tissue paper. Leave at least 1 cm on the side of the strip of tissue paper so you can tape it to a sheet of white paper towel.

2. Tape the other strip of tissue paper next to the strip containing the petroleum jelly. With a pencil, write PETROLEUM and CONTROL under the appropriate tissue paper strip on the paper towel.

3. Tape the paper towel to a piece of cardboard that is slightly larger than the sheet of paper towel. Write your name at the bottom of the piece of cardboard.

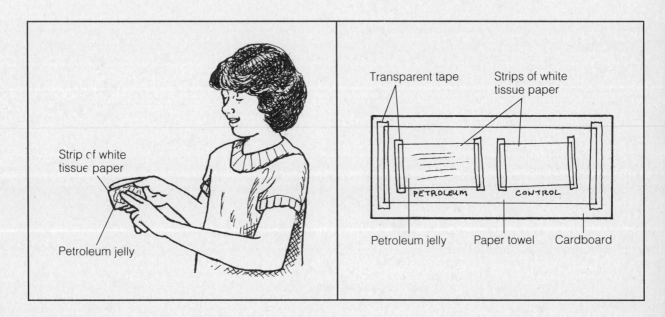

© Prentice-Hall, Inc.

Human Biology and Health H ■ 239

4. Place several drops of food coloring into a plastic container that is half filled with water. Stir the colored water with a spoon.
5. Place one drop of colored water in the middle of each tissue paper strip. Put the cardboard in an area where it will remain undisturbed for one day.

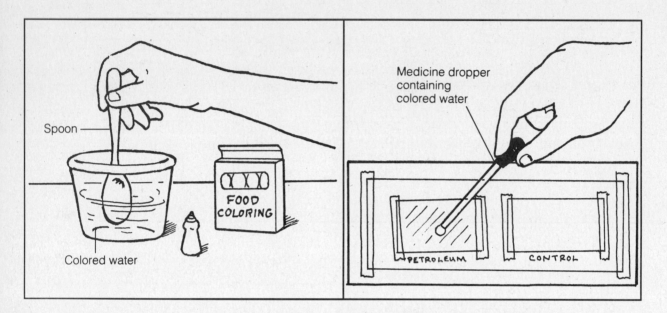

6. After one day, remove the tissue paper strips from the paper towel. Examine the paper towel.
7. Share your results with your classmates.

Observations

Describe what you observed when you removed the tissue paper. _____
_____
_____
_____
_____

Analysis and Conclusions

1. What effect does petroleum jelly have on water? _____
_____
_____

2. Why was a control included? _____
_____
_____

Name _____ Class _____ Date _____

3. What does this activity tell you about petroleum jelly? _____
_____
_____
_____

4. How did your results compare with those of your classmates? _____
_____
_____
_____
_____

5. What other substances may work as well as petroleum jelly? _____
_____
_____
_____

Activity

Respiratory and Excretory Systems

CHAPTER 5

How Fast Do Your Nails Grow?

Your nails are rough and are made of a protein called keratin that gives them their characteristic strength. The visible part of the nail is called the body, or nail plate. It grows out of the root, which is hidden beneath the nail at its base. Also at the base, there is a whitish half-moon shape called the lunula (lunula is Latin for little moon). The area under the nail is called the nail bed. To find out how long it takes for your nail to grow from its base to its top, try this activity.

What You Will Need
thin paintbrush
acrylic paint
metric ruler

What You Will Do
1. Using a thin paintbrush, place a tiny dot of acrylic paint above the cuticle (area of hardened skin) at the base of one of your fingernails and one of your toenails. Let the paint dry. Allow the acrylic paint to remain there undisturbed for three weeks.

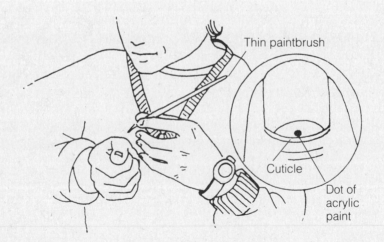

2. Observe the acrylic-paint dots each day. If they begin to wear away, put another dot of paint on. Be sure to place the new acrylic-paint dot exactly on top of the old acrylic-paint dot.

3. At the end of each week, use a metric ruler to measure the distance from the cuticle to the acrylic-paint dot on each nail. Record this measurement in the Data Table on the next page. Continue to do this until the dot grows to the point that you cut it off when you clip your nails.

© Prentice-Hall, Inc.   Human Biology and Health H ■ 243

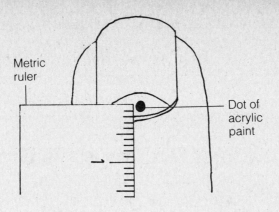

What Did You Find Out?

DATA TABLE

Day	Distance from Cuticle	
	Fingernail	Toenail
1		
7		
14		
21		

1. What was the average weekly growth of your fingernail? Your toenail?

2. Which nail grew faster—fingernail or toenail? Suggest a possible explanation for your observation.

3. Compare your results with those of your classmates.

4. What functions do your nails serve?

Activity

Nervous and Endocrine Systems

CHAPTER 6

How Fast Can You React?

Imagine you are riding your bicycle in the park. Suddenly a squirrel darts out across the path in front of you. You react quickly and miss hitting the squirrel by centimeters. The length of time that passed between your seeing a change in your environment (squirrel darting out in front of you) and your reacting to that change (turning the bicycle away) is called your reaction time. Why is it important to react quickly to situations such as this? Can you measure your reaction time? By doing the following activity you will find out the answers to these questions. All that you will need for this activity is a metric ruler.

Procedure
1. Have your partner hold a metric ruler vertically about 50 cm above a table.
2. Without touching the ruler, position your thumb and forefinger around the zero mark. See the diagram below.

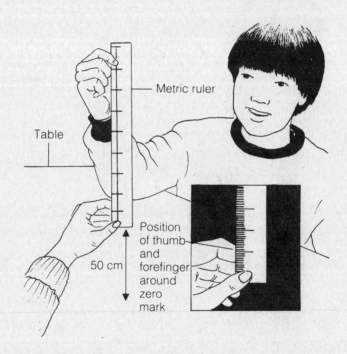

3. Your partner should drop the ruler whenever he or she chooses. Moving only your thumb and forefinger (not your hand), try to catch the ruler as soon as it falls.

4. In the Data Table below, record the distance in centimeters that the ruler falls.
5. Repeat steps 1 through 4 four more times.
6. To obtain an average distance, add the five distances together and divide the sum by five. Record the average distance in the Data Table.
7. Reverse roles with your partner and repeat steps 1 through 6.

Observations

DATA TABLE

Trial	Distance Metric Ruler Falls (cm)
1	
2	
3	
4	
5	
Average	

Name _____ Class _____ Date _____

Analysis and Conclusions

1. Why does measuring the distance that the ruler falls give a relative measure of reaction time?

   _____
   _____
   _____
   _____

2. Why is it important to calculate an average reaction time?

   _____
   _____
   _____
   _____

3. Compare the reaction times of all your classmates. What can you conclude about all of the reaction times?

   _____
   _____
   _____
   _____
   _____
   _____

Going Further

Design an investigation in which you determine the effect fatigue has on your reaction time.

© Prentice-Hall, Inc.

Human Biology and Health   H ■ 247

Activity

Nervous and Endocrine Systems

CHAPTER 6

A Gentle Touch

If a large insect crawls on you, you can usually tell where it is, even with your eyes closed. The reason is that whenever something touches you, or you touch something, your sense of touch goes into action. To discover the remarkable things your sense of touch is capable of, try this activity.

Things You Will Need
nickel
dime
penny
quarter
small shoe box with cover
blindfold

Directions
1. Place the nickel, dime, penny, and quarter in the shoe box and cover the box.
2. Blindfold your partner and have your partner reach into the shoe box and remove one coin.
3. Have your partner identify the coin by holding it in his or her hand and touching it.

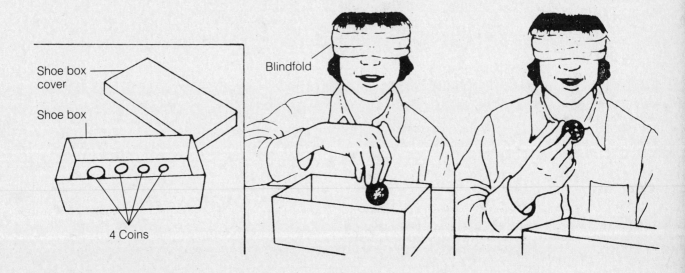

4. After your partner has identified the coin, have your partner remove the blindfold to see if he or she was correct.
5. Repeat steps 2 through 4 three more times. Record your observations in the Data Table on the next page.
6. Remove the coins from the shoe box and set them flat on a table.

© Prentice-Hall, Inc.

Human Biology and Health H ■ 249

7. Put the blindfold on your partner again and this time have your partner touch the top of each coin with only the tip of his or her index finger. **Note:** *Do not allow your partner to pick up the coins.*

8. Have your partner identify each coin.
9. Repeat steps 7 and 8 three more times. Record your observations in the Data Table.
10. Reverse roles with your partner and repeat steps 1 through 9.

Things to Look For

DATA TABLE

Trial	Holding Coin in Hand		Touching Coin With Fingertip	
	What I Think Coin is	What Coin Actually Is	What I Think Coin Is	What Coin Actually Is
1				
2				
3				
4				

Conclusions to Draw

1. Was it easier to identify the coins by picking them up and touching them or by touching their tops with your index finger? Give a reason for your answer.

_____
_____
_____
_____

Name _____ Class _____ Date _____

2. In which case were you actually using only touch receptors to identify the coins?
   _____
   _____
   _____

3. In addition to touch receptors, what other receptors might you be using?
   _____
   _____
   _____

4. Compare your results with those of your classmates.
   _____
   _____
   _____

Name _____ Class _____ Date _____

Activity
Nervous and Endocrine Systems
CHAPTER 6

Colored "Sandwiches"

Why is a stoplight red, grass green, and an asphalt road black? The answer to this question depends on the nature of light and the colors of light striking the object. Light is a form of energy that behaves in some ways like waves. Light waves have a range of wavelengths. The different wavelengths of light appear to you as different colors—red, orange, yellow, green, blue, indigo, and violet. Light that contains all of these wavelengths is called white light. When light strikes the surface of an object, the light can be transmitted, reflected, or absorbed.

If an object such as one of the colored "sandwiches" you will construct allows light to pass through it, the color that you see is the color that reaches your eyes. The other colors are absorbed.

Try your hand at the following activity to see if you can predict what color you will see when you look at a source of white light (sunlight) through colored "sandwiches" like those shown in the drawing below.

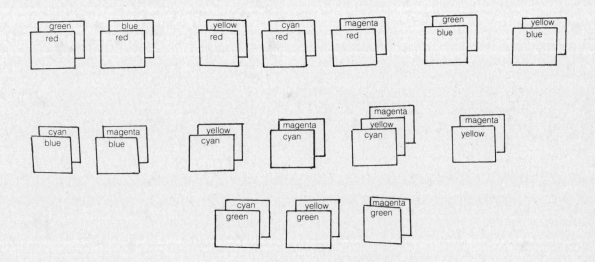

Here is a list of the materials that you will need for this activity.

brightly lighted window
sheet of unlined white paper
red, green, yellow, magenta (purplish red),
   cyan (dark blue) colored plastic or glass filters

© Prentice-Hall, Inc.   Human Biology and Health   H ■ 253

Before you actually perform the activity, place your prediction next to the appropriate "sandwich" in the Data Table. Now use the pieces of colored plastic or glass filters to make your own "sandwiches" and find out if you were correct. Allow the light from a brightly lighted window to pass through your "sandwiches" onto a sheet of white paper. Stand about one meter away from the lighted window. Record the actual color that you see on the paper in the Data Table.

DATA TABLE

"Sandwich"	Predicted Color	Actual Color
red/green		
red/blue		
red/yellow		
red/cyan		
red/magenta		
blue/green		
blue/yellow		
blue/cyan		
blue/magenta		
cyan/yellow		
cyan/magenta		
cyan/yellow/magenta		
yellow/magenta		
green/cyan		
green/yellow		
green/magenta		

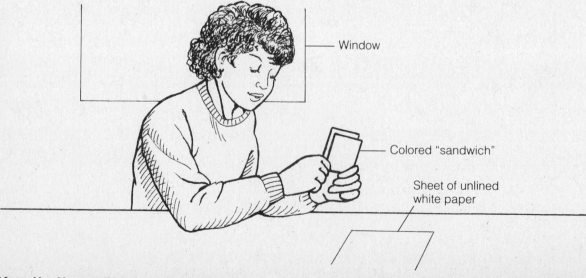

Name _____ Class _____ Date _____

Now see if you can answer these questions.

1. Were your predictions correct? _____
   _____
   _____

2. What explanation can you give if your predictions were not entirely correct?
   _____
   _____
   _____
   _____

3. Why is ordinary window glass said to be colorless? _____
   _____
   _____
   _____

4. How did your results compare with your classmates? Can you explain any differences in results? _____
   _____
   _____
   _____
   _____
   _____

Name _____ Class _____ Date _____

Activity

Nervous and Endocrine Systems

CHAPTER 6

More Than Meets the Eye

Look at the drawing below. Which center dot is bigger in these two sets of circles? The correct answer is that both center dots are the same size.

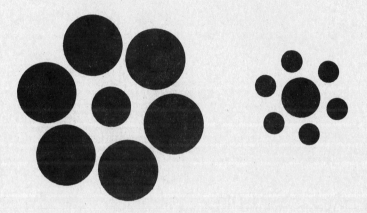

Sometimes your eyes play tricks on you. The way you see an image may be different from its true shape, size, color, or movement. This is called an optical illusion. Look at the following drawings and see if you can tell what is real and what is an illusion.

1. What do you see at the intersections of the boxes shown above? _____
_____
_____

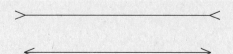

2. Which line is longer? _____

© Prentice-Hall, Inc. Human Biology and Health H ■ 257

3. What do you see in the drawing? _____
_____
_____

Name _____ Class _____ Date _____

Activity

Nervous and Endocrine Systems

CHAPTER 6

Some Sound Reasons

Sound is a form of energy with some very interesting characteristics. Why not try these activities and find out how interesting sound really is.

Part A: Can You See Sound?

A sound is produced when matter vibrates and the vibrations travel as a wave through a medium. Can you see the vibrations? To answer this question, gather these materials together—a tuning fork, a pencil, and a glass of water—and follow the directions below.

1. Strike the prongs of a tuning fork against the pencil or the heel of your shoe, then hold the fork close to your ear.

    What happens? _____
    _____
    _____

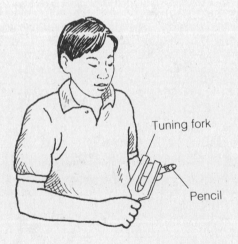

2. What happens when you touch the prongs of the fork? _____
    _____
    _____

© Prentice-Hall, Inc.

Human Biology and Health H ■ 259

3. Again strike the prongs of a tuning fork against the pencil or the heel of your shoe and then place the ends of the prongs in a glass of water.

   What happens? _____
   _____
   _____

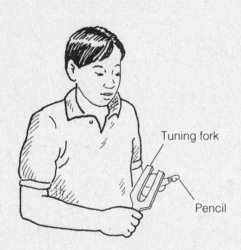

Part B: Can Sound Turn a Corner?
1. Have your partner walk around a corner. Then stand on the other side of the corner, face straight ahead, and call your partner's name.

   Does he or she hear you? _____

Name _____ Class _____ Date _____

2. Now standing in the same place and facing the same way, try to throw a ball to your partner.

    Does he or she catch it? _____

3. How might you explain your observations? _____
_____
_____
_____

Part C: Why Do You Hear Different Sounds?

How you hear and describe a sound depends upon the physical characteristics of the sound waves. For example, certain sounds are described as high—such as those produced by a piccolo—or low—such as those produced by a tuba. A description of a sound as high or low is known as the pitch of a sound. The pitch of a sound depends on the frequency of waves.

1. Strike the prongs of two tuning forks of different frequency against the pencil or the heel of your shoe. Bring one of the tuning forks nearer to your ear. Then bring the other tuning fork nearer to your ear.

    Is there a difference in sound? _____

2. Does one tuning fork have a higher pitch than the other? Which tuning fork has the higher pitch? _____
_____
_____

© Prentice-Hall, Inc.   Human Biology and Health

3. Strike one of the tuning forks against the pencil or your heel. Then place the ends of the prongs in a glass of water and notice the wave vibrations in the water. Repeat this action with the other tuning fork.

Was there a difference in the wave vibrations? _____
_____
_____

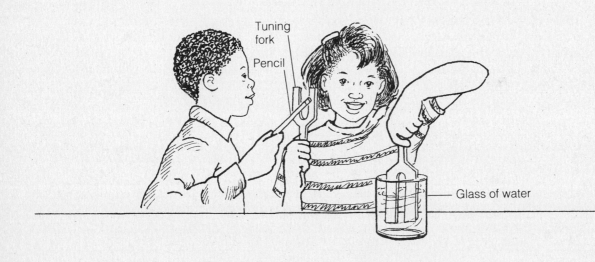

4. What is the relationship between pitch and frequency of wave vibrations? _____
_____
_____
_____
_____
_____

262 ■ H   Human Biology and Health

Name _____ Class _____ Date _____

Activity

Immune System

CHAPTER 8

It's No Skin Off Your Nose

The skin is the body's largest organ. Its tough outer covering protects the body from invading microorganisms. As long as the skin remains uninjured, the invading microorganisms are kept outside the body. But what happens when the skin is injured? Find out by doing this activity.

Materials
4 sheets of unlined white paper
pencil
soap
4 apples
straight pin
cotton swab
rubbing alcohol

Procedure

1. Place four sheets of paper on a flat surface. Label the papers A, B, C, and D. Carefully wash your hands with soap and water. Dry them thoroughly.

2. Wash four apples and place one apple on each sheet of paper. Do not touch apple D for the remainder of this activity.

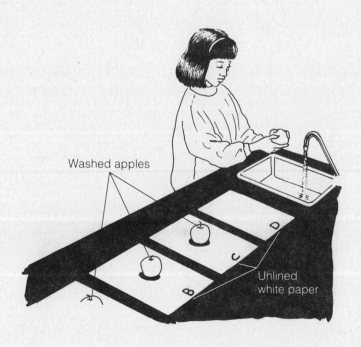

© Prentice-Hall, Inc.

Human Biology and Health H ■ 263

3. Using a straight pin, puncture four holes in apple B and four holes in apple C.

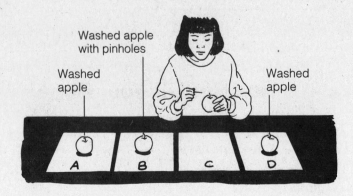

4. Have your partner who has not washed his or her hands hold and rub apple A, apple B, and apple C.

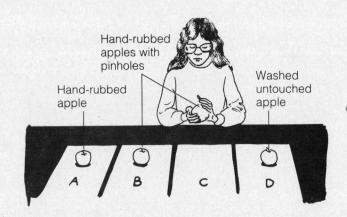

5. Dip a cotton swab in rubbing alcohol. **CAUTION:** *Be careful when using rubbing alcohol.* Apply the alcohol to the punctured areas of apple B.

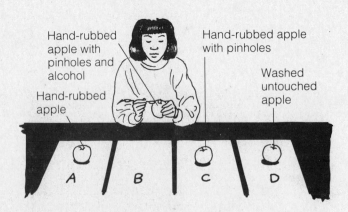

Name _____ Class _____ Date _____

6. Place the four apples in an area where they will remain undisturbed for one week. Examine each apple every day. **Note:** *Do not touch the apples while examining them.* Arrange your observations in the Data Table.
7. After one week, compare the apples.
8. Share your results with the class.

Observations
DATA TABLE

Day	Observations
1	
2	
3	

Analysis and Conclusions

1. How did the apples compare? _____
_____
_____
_____
_____

2. What was the purpose of apple D? _____
_____
_____

© Prentice-Hall, Inc.   Human Biology and Health   H ■ 265

3. What is the relationship between the apples and your skin? _____

_____

_____

_____

4. How did your results compare with those of your classmates? Were they similar? Different? Explain any differences. _____

_____

_____

_____

_____

_____

_____

Answer Key

ACTIVITY BANK

Activity: A Human Cell vs. an Ameba

Observations Students should observe a cell membrane and a nucleus in both the human cheek cell and the ameba. In the ameba, they should also see vacuoles. **Analysis and Conclusions 1.** The human cheek cell and the ameba are similar in that they both contain a cell membrane, a nucleus, and cytoplasm. **2.** They are different in that the ameba has more vacuoles and is a living organism able to move, get its food, and so on. The human cheek cell, although it comes from a living organism, is only a very small part of that organism and once it is removed from the organism, it will die. **3.** Humans are made of trillions of cells; the ameba is made of one cell. **4.** Student results should be similar.

Activity: Under Tension

What You Will Discover 1. The hairpin "walked" along the knife's edge. This action shows that there are small movements in the muscles of the arm, which cause the knife's edge to move up and down slightly and the hairpin to "walk." **2.** The hairpin "walked" along the knife's edge faster. The tighter the arm was held, the more the muscles contracted. **3.** Students' results should be similar to the results of the individual student.

Activity: Getting the Iron Out

Observations Grayish, metal particles. **Analysis and Conclusions 1.** Metal (iron). **2.** A mixture is a combination of two or more substances that are not chemically combined. The substances in the cereal have kept their separate identities and most of their own properties. **3.** It is possible to separate the metal particles from the cereal because the cereal is a mixture. In a mixture, all the methods used to separate substances are based on the physical properties of the substances making up the mixture. In this case, the iron, because it has magnetic properties, was removed from the cereal using a magnet. **4.** The cereal is a mixture made up of different materials unevenly spread out (heterogeneous mixture).

Activity: Going Crackers

Observations Test tube 1 was tested for starch and it was found to be present. Test tube 2 was tested for sugar and it was found to be absent. Test tube 3 was tested for starch and it was found to be absent or there was very little present. Test tube 4 was tested for sugar and it was present. **Analysis and Conclusions 1.** Iodine solution was added to test for the presence of starch. If starch is present, the mixture will turn blue-black. **2.** Benedict's solution was added to test tubes 2 and 4 to test for the presence of sugar. If the mixture turns green, orange, or red after heating, sugar is present. **3.** Before chewing, the soda cracker tasted bland and dry. After chewing for 5 minutes, the soda cracker tasted sweet. **4.** The starch in the soda cracker was broken down (digested) into sugar by the enzyme ptyalin, which is present in saliva. **5.** In the mouth, starch is broken down into sugar by ptyalin. **6.** Student answers will be similar, unless the soda cracker was not chewed long enough.

Activity: Do You Have the Stomach for . . . ?

Teacher Note Use a 0.2% solution of hydrochloric acid for this activity.
Observations DATA TABLE 1 At the start, the egg white cubes will be the same size and color in all the test tubes. After one day, however, the egg white cubes in test tube B will seem to disappear, leaving a liquidy mass. There is hardly a change in the appearance and size of the egg white cubes in test tubes A and C after one day. **DATA TABLE 2** The materials in test tubes A and C will not cause a change in blue litmus paper (it will remain blue) both at the start of the activity and after one day. The material in test tube B will turn blue litmus paper pink

© Prentice-Hall, Inc.

Human Biology and Health   H ■ 267

both at the start of the activity and after one day. This is because of the presence of hydrochloric acid. **Analysis and Conclusions 1.** The addition of hydrochloric acid caused the egg white (protein) to break down (digest), leaving a soft, liquidy mass. **2.** To allow the substance(s) time to react. **3.** Pepsin and hydrochloric acid (test tube B); pepsin and water (test tube C); The presence of hydrochloric acid will greatly speed up the digestion of proteins. **4.** Pepsin digests protein in the stomach in an acid environment. **5.** Student results should be similar. Differences in results may occur if the egg white cubes are too large or if the solution of hydrochloric acid is too weak. See Teacher Note.

Activity: A Pulsating Question

What Did You See? The average pulse rate of an earthworm is 25 to 30 times a minute. The average pulse rate of a human is 60 to 80 times a minute. **What Did You Discover? 1.** To get a more accurate indication of a pulse rate, it is better to take an average of the pulse rate. This is because a pulse rate can change from minute to minute. **2.** The earthworm's pulse rate is slower than the student's pulse rate. **3.** In addition to the obvious animal characteristics, humans and earthworms also have closed circulatory systems (blood circulates within blood vessels). The blood of each is also pushed throughout the system by a pumping mechanism. In humans, it is the heart; in earthworms, it is a series of 5 "hearts," or aortic arches. **4.** Student results should be similar. However, changes in temperature and pH can affect the pulse rate of the earthworm.

Activity: You've Got to Have Heart

What You Will Discover 1. The action of squeezing the bottle with two hands squirted water farther. The use of two hands causes a greater pressure to be exerted on the bottle, thereby forcing the water inside to be forced out through the tube with a greater force. **2.** The action of squeezing the bottle with two hands better simulates the action of the heart. This is because the heart has two separate pumping chambers. **3.** The greater the force exerted on the liquid, the greater its pressure and therefore the greater distance it travels. **4.** When blood is pumped out of the heart it travels through arteries and then through capillaries. **5.** Student results should be similar.

Activity: The Squeeze is On

What Did You See? Water should squirt farther from the glass tube. **What Did You Discover? 1.** Artery— glass tube; vein—plastic tube. **2.** Blood. **3.** Blood pressure is higher in arteries than in veins because the arteries have thick, muscular walls that stretch only a little each time the heart pumps blood in them. **4.** Heart. **5.** Student results should be similar.

Activity: Taking a Breather

Teacher Note You may wish to demonstrate the procedure for exhaling into the limewater and to show students what cloudy limewater looks like. **What Did You See?** Students will observe that the limewater turns cloudy in less time after exercise than it does at rest. They will also observe that the number of breaths increased after exercise. **What Did You Discover? 1.** The limewater turns cloudy faster after exercise. **2.** The lungs work harder while exercising because the body needs more oxygen and needs to get it into the body faster to do the exercise. As a result, more of the waste gas carbon dioxide is produced. This is supported by the fact that the limewater turns cloudy faster after exercising than it does at rest. **3.** There is more carbon dioxide in your body during exercise than there is at rest because the body cells need more energy. In order to get the energy, the process of respiration is speeded up, thereby causing an increase in the amount of waste products—carbon doxide and water—that are produced. **4.** Student results should be similar. **5.** Breathing would be more labored because the air at higher altitudes does not contain as much oxygen as it does at lower altitudes. Thus, a person would tire quickly and would not be able to do most of the activities that he or she could do at lower altitudes.

Activity: Lip Service?

Observations There was a colored stain on the paper towel directly under the tissue paper strip that contained no petroleum jelly. There was no stain on the paper towel directly under the tissue paper strip containing the petroleum jelly. **Analysis**

and Conclusions 1. The petroleum jelly blocks the passage of water. 2. To show what would happen without applying the petroleum jelly. 3. Petroleum jelly blocks the passage of water. 4. Student results should be similar. 5. Lip balms, lipsticks, and so on.

Activity: How Fast Do Your Nails Grow?

What Did You Find Out? 1. 0.16 cm/week. 2. Fingernail. Fingernails grow more quickly than toenails because fingernails perform more activities that involve putting pressure on the fingernails. 3. Results should be consistent with those of individual students. 4. Nails protect the sensitive upper layers of the tips of fingers and toes. With these surfaces protected, the fingers become more useful in grasping objects and in prodding and scratching. Toenails, covering the tips of the toes, protect the toes when walking or running.

Activity: How Fast Can You React?

Observations Answers will vary. **Analysis and Conclusions** 1. It gave a relative measure of the amount of time it took the person to react to a particular controlled situation. It did not give an absolute measure of reaction time in general but only in terms of catching the meterstick. 2. To obtain an average reaction time. During a single trial, other variables could come into play. 3. Answers will vary. Each student is an individual with his or her own particular coordination, reaction time, and so on.

Activity: A Gentle Touch

Things to Look For Answers will vary. **Conclusions to Draw** 1. It was easier to identify the coins by picking them up and touching them. Manipulation of the coins leads to perceptions of pressure, movement, resistance, and position. These perceptions enable the brain to develop a three-dimensional re-creation of the object, which permits an easier identification. 2. When the coins were touched with the index finger only. 3. Pressure. 4. Student results should be similar.

Activity: Colored "Sandwiches"

1. and 2. Student predictions may not be quite right. Most filters let through small amounts of some colors that are not expected. For example, some blue filters transmit a small amount of red light. Consequently, although students would predict that no light would come through a red/blue "sandwich," they may see a faint red color through the "sandwich" because the filters are not perfect. The results, thus, would depend on the quality of the colored filters being used. 3. Ordinary window glass is said to be colorless because it can transmit all colors. 4. See answer to 1 and 2.

Activity: More Than Meets the Eye

Teacher Note The messages sent by the optic nerves to the brain resemble the television signal sent from a TV station. In putting the picture together, the brain has to interpret the incoming messages from the rods and cones of the eye. Most of the time, the brain can be very accurate. Sometimes, however, it can be fooled by puzzling clues. This is exactly what happens when you look at optical illusions. 1. Persistent gray spots, which are not really there—they are only in your "mind." 2. Both lines are the same length. 3. A goblet and/or two profiles facing each other.

Activity: Some Sound Reasons

Part A 1. The tuning fork vibrates and gives off sound. 2. When the prongs are touched, the vibrations can be felt. 3. When the vibrating prongs are placed in water, they cause the water to splash. **Part B** 1. Yes. 2. No. 3. When the ball is thrown, it moves in a straight line, not around the corner. Although sound too travels in a straight line through the air, because it is a wave it can bend around an encountered obstacle and pass into the area behind the obstacle. **Part C** 1. Yes. 2. Yes; the tuning fork that vibrates faster (has the higher frequency) will have the higher pitch. 3. Yes; the tuning fork with the higher pitch will vibrate faster and create more spashing in water. 4. Pitch is a property of sound that depends on the frequency of waves. Sound waves that have a high frequency are heard as sounds of high pitch. Sound waves that have a low frequency are heard as sound of low pitch.

Activity: It's No Skin Off Your Nose

Observations Answers may vary. **Analysis and Conclusions 1.** Apples A and D should not show any change. There should be decay where the punctures were made in apple C. Apple B may show decay where the punctures were made, but to a much lesser degree than apple C. **2.** To serve as a control. **3.** The apple skin protects the apple in much the same way as the skin protects the inside of the body. This should be evident in comparing apples A and C. When foreign substances penetrate the skin, such as through an open wound, the importance of using an antiseptic should be evident by comparing apples B and C. **4.** Student results should be similar.